Political Ecology in a
Yucatec Maya Community

Political Ecology in a Yucatec Maya Community

E. N. Anderson

with

Aurora Dzib Xihum de Cen
Felix Medina Tzuc
Pastor Valdez Chale

THE UNIVERSITY OF ARIZONA PRESS
TUCSON

The University of Arizona Press
© 2005 Arizona Board of Regents
All rights reserved
∞ This book is printed on acid-free, archival-quality paper.
Manufactured in the United States of America
10   09   08   07   06   05   6   5   4   3   2   1

Library of Congress Cataloging-in-Publication Data

Political ecology in a Yucatec Maya community / E.N. Anderson
... [et al.].
       p. cm.
   Includes bibliographical references and index.
   ISBN-13: 978-0-8165-2393-1 (hardcover : alk. paper)
   ISBN-10: 0-8165-2393-2 (hardcover : alk. paper)
   1. Mayas—Agriculture. 2. Mayas—Land tenure. 3. Mayas—
Medicine. 4. Indigenous peoples—Ecology—Mexico—
Chunhuhub. 5. Land use—Mexico—Chunhuhub. 6. Logging—
Mexico—Chunhuhub. 7. Biotic communities—Mexico—
Chunhuhub. 8. Conservation of natural resources—Mexico—
Chunhuhub. 9. Chunhuhub (Mexico)—Environmental
conditions. 10. Chunhuhub (Mexico)—Social life and customs.
I. Anderson, Eugene N. (Eugene Newton), 1941–
F1435.3.A37P65 2005
305.897'42–dc22

                                            2005015644

For the children of Quintana Roo
And all the children

Contents

Illustrations

All photos by E. N. Anderson except as otherwise noted in the individual photo captions.

Preface

THIS BOOK is a study in political and cultural ecology—how people manage their lands and resources. In that enterprise, the Maya of the Yucatán Peninsula have had more than their share of successes, and some failures as well. They have much to teach about tropical development and about the human condition.

Chunhuhub, "Town of the Wild Plum Trees,"[1] is a community of some six thousand people, almost all of them Yucatec Maya, in the western part of the "Maya zone" of Quintana Roo, Mexico's southeastern state. Since 1990, I have been more and more closely involved with the life of Chunhuhub and its small satellite towns, especially Presidente Juarez, a few kilometers to the southwest. This book provides a portrait of Chunhuhub as it enters the twenty-first century.

The region lives by farming. I have focused, therefore, on the rural economy—on Maya knowledge of plants and animals, and how that knowledge affects management. This has led me to look at the history, politics, and everyday life of the town. Farming and forest use cannot be separated from other aspects of life in the community. Quite apart from the "holism" that is a matter of faith among anthropologists (at least those of my generation), the pragmatics of plant and animal use in farming regions are so tightly enmeshed in all aspects of life that they cannot be understood without at least some reference to history, kinship, religion, education, and political economy.

Three people—my field assistants, Felix Medina Tzuc and Aurora Dzib, and my landlord and friend, Pastor Valdez—are present, in one way or another, on every page. Their input demands recognition, but they do not wish to be associated with my theorizing and worldwide perspective, hence the ambiguous "with" in the author line. Much of the message in this book derives from conversations with them and our various neighbors. Such spontaneous conversations do not lend themselves to quantification.

The purpose of this book is not to resolve all the questions; it would be the height of arrogance for me to claim the ability to resolve even one of them. I have spent a total of a year and a half in the Yucatán Peninsula, most of it in one small town. I have talked to relatively few people.

I have, instead, learned a good deal—if only a fraction of what there is to learn—about Maya rural economy. I have listened to many conversations about the environment, how to manage it, and what to do now that changes are rapid and disorienting. I have attended sessions in which key decisions were made about resource use and policy. I have interviewed people about ecotourism, conservation, and harvest of wild resources. I have watched farming activities and censused gardens and vegetable plots. I am, therefore, in a position to transmit many ideas and questions, and to ask a few questions of my own.

Denise Brown, now director of Latin American Studies at the University of Calgary, surveyed communities around the Yucatán Peninsula and told me Chunhuhub would be ideal for my work. I had learned to trust her judgment, and, indeed, she was right.

In Chunhuhub, I reside with the family of Pastor Valdez and his wife, Zenaida Estrella de Valdez, from whom I rent a room. Don Pastor is a bilingual teacher and the principal of a school in Presidente Juarez. (Adult heads of households are Don or Doña in the Zona Maya of Quintana Roo: "Don Pastor" and "Doña Zenaida," not "Sr. y Sra. Valdez.") They possess a modernized Maya house compound—two cement-block dwellings on a small lot in the center of town.

In 1991, I was there for five months in the village: February through June. There was a small stove, an unusual luxury in Chunhuhub. As in almost all Chunhuhub houses, electricity and running water were available, if not always reliable. Sleeping facilities were the hammocks typical of the area, and bathing, laundering, and so on were of traditional type. Foods in Chunhuhub were those of the area: tortillas, black beans, locally

Right to left, Felix Medina Tzuc,
Pastor Valdez, and E. N. Anderson
on the great pyramid at Altamirano;
photograph by Betty Faust.

Aurora Dzib
Xihum de Cen
with achiote bush,
Presidente Juarez.

baked breads, pork, the limited range of local vegetables and fruits, and occasionally poultry or game.

My wife and I were more or less "participant observers." We were atypical in having an automobile and drinking bottled water. The local water supply was unsafe to the inexperienced, unless boiled for a period of time that was quite unconscionable, given the family's need for their stove. Since 1991, many families have shifted to using bottled drinking water.

In 1996 I was there alone for six months (from January to mid-June), and in 2001 for three (September through November); in November, a student, Tracy Franklin, was also there, researching midwifery (B. Anderson et al. 2004). I have made short visits to the town in almost every year since 1991. Brief visits by Eugene Hunn and Betty Faust also produced important new information.

The nature of fieldwork introduces its own biases. Inevitably, I know the world of men in their fifties—grandfathers, heads of households—better than the worlds of the young and the female.

The people of Chunhuhub were invariably hospitable, friendly, and polite. In years of usually blissful fieldwork in several countries, I have never had a more uniformly warm and accepting field situation. A few people suspected my motives (or sanity), but they continued to obey the community norms of hospitality and politeness. I was suspected by some of wanting to kidnap children and take their eyes. This I took as the wildest of rumors, but in fact Maya children were being kidnapped in neighboring Guatemala and killed for their body parts (Scott Atran, personal communication, 2002). In any case, friends rallied to my defense, and quickly caused such rumors to be dismissed.

My fellow field-workers and I, on our part, were careful not to ask too many questions or inquire about possibly embarrassing matters. This doubtless cost us some valuable information, but on the other hand we had none of the confrontations or misunderstandings that make up the stuff of fieldwork confessional literature.

To offset the natural tendency to romanticize the place, I have tried to confine myself to clearly verifiable facts (the number of thatched houses, the price of cattle) or to opinions I recorded from at least a few citizens. I have expressed opinions, but I have tried to keep them relevant to wider concerns, where my local biases are less relevant and where an informed personal judgment is, arguably, a desirable thing.

When I look at Maya life, in all its diversity, I tend to see it from a distinctively Chunhuhubian point of view. There are other view-stations. Eric Villanueva Mukul is the latest (and perhaps the most incisive) of many who have looked from Mérida across the seared henequen lands around that city (Villanueva Mukul 1985, 1990). A large group centered on the University of Yucatán's botany faculty has worked mostly in the fertile but crowded south of Yucatán state, as have INAH (Instituto Nacional de Antropología e Historia) investigators such as Carmen Morales Valderrama (1987). Other investigators have worked in the eastern part of the Zona Maya of Quintana Roo, especially around the old centers of Señor and Tixcacal Guardia; Villa Rojas (1945, 1978) was first, and paid the price of appearing in a less than ideal light in some of Paul Sullivan's "unfinished conversations" (Sullivan 1989). More recently, Mexican, Canadian, and U.S. workers have filled many blank spaces on the map.[2]

Recently, major moral questions have been raised about the advisability of publishing ethnobotanical data (Anderson 2001; Shiva 1997; Vogel 2000). The freewheeling appropriation of genetic material and cultural information is no longer acceptable. Even full compensation according to reasonable current standards is no longer adequate, since there is a very slight chance that enormous wealth could follow from the development of a plant newly found, and then the compensation given would no longer be "reasonable."

I have left all specimens in cooperating agencies in Yucatán and Quintana Roo. All specimens are deposited in the herbaria of the University of Yucatán (Mérida) and ECOSUR (El Colegio de la Frontera Sur, Chetumal). I compensated all consultants at an appropriate rate, especially those who were professionals and deserved professional fees. I relied, as so often, on Don José Cauich Canul, the healer of the neighboring town of Polyuc, for help with this. As a practitioner with widespread experience and a large urban clientele, he knows what his services are worth, and I paid accordingly. This provided a guideline for compensation to others. The knowledge herein presented was not obtained through piracy of any sort. The only problem that could ensue would be the discovery of an enormously successful new drug or food from these materials. Then, there would arise the question of capturing for the Maya the appropriate revenues from this innovation.

A vast literature describes Maya plant use (esp. Anderson 2003; Ankli 2000; Ankli et al. 1999a, 1999b; Barrera Marín et al. 1976; Herrera Castro 1994; Pulido Salas and Serralta Peraza 1993; Sosa et al. 1985). Moreover, most of the plants in question are found widely in Mexico and the Caribbean, and thus have many uses recorded in standard compilations from around the region (esp. Morton 1981). There is little point in hiding information already in the public domain or at least easily accessible. Otherwise, thanks to consultation with Joseph Vogel, I have decided to abstain from even providing a sense of what I am not revealing. People I work with have signed releases for me to use and publish their knowledge, but there is a very real question of whether this is "informed" consent in the case of those who are monolingual and far removed from the world of scientific publication. Fortunately, this does not apply to the three cited above and on the title page, or to Don José Cauich Canul; all are literate and quite familiar with the wider world. Many of the Yucatec Maya are more or less aware of the problems with ethnobotanical publication that have occurred among the Highland Maya of Chiapas, so consent is informed in these cases. All major consultants have encouraged me to publish. Don Pastor, in particular, always asks me for the results of my work in an account that will be useful to the community. Also, in 2001 and 2002, the mayor and *comisario ejidal* (head of the landholding community), while signing permissions on behalf of the community, both made me promise to get my information out!

I have not worked with drug companies or made information available to them. The current state of play makes any joint ventures a minefield. We all hope the Maya will commercialize their own drugs, as they have begun to do in many areas. We hope to be able to aid such ventures.

Considerations of space and reader patience have made me leave out a great deal of information, ranging from a list of occupations in Chunhuhub to a review of the literature on Quintana Roo Maya communities. This information is available from me on request.

Several points regarding terminology in this book should be clarified. First, money is given in pesos. In 1991, the peso was running about 10,000 to the U.S. dollar. The new peso, worth 1,000 old pesos, was introduced soon after, and the rate has been roughly 10 to the dollar since.

The Yucatán Peninsula is a large area that includes all of southeast Mexico, the northern end of Guatemala, and the independent nation of Belize. Southeast Mexico is divided into three states: Campeche, Yucatán, and Quintana Roo. Yucatán is by far the most populous of these. This introduces a terminological problem. Many authors use "Yucatán" to mean the whole peninsula. I cannot reasonably do this, since I have to speak repeatedly of Yucatán state, the Mexican political division that includes the northern and northwestern sides of the peninsula (Quintana Roo is the eastern side). I refer to the whole region as "the Yucatán Peninsula," or "the peninsula." I refer to the political entity as "Yucatán state."

Since the Spanish came, the Yucatán Peninsula has been the Land of Misunderstanding. Spanish landing in Campeche (so the story goes; Restall 1998:122 cites an eighteenth-century Maya source) asked local people the name of the land. This must have drawn a blank stare from the person addressed, because some helpful soul said, "He didn't understand you" (written *matan cub a than* in Restall's source; see chapter 5). This story is also recorded by William Prescott (1902 [1843], 165–66) in *The Conquest of Mexico*. He gives the phrase as *tectetan*, which makes no sense, but notes that the Mayanist Baron Waldeck corrected it to "*Ouyouckatan*" (i.e., the phrase I cite, in French spelling and without the negative particles). It literally means "he didn't hear what you said," but idiomatically that means "he didn't understand you." The Spanish were content to take the accented syllables as the name of the place. This was the first, but not the last, misunderstanding in the history of Spanish-Maya relations (Frazier 1997).

Confusion is compounded by the anthropologists' use of the term "Yucatec Maya" to refer to the people who call themselves "Maya" (more accurately, *Maayaj*) and call their language the "Maya" language (*maayaj t'aan*). There are well over twenty related languages spoken in south Mexico, Guatemala, Belize, and Honduras. Quite early, scholars began to refer to these as "Maya languages." That meant that someone had to generate a term to set off the original "Maya language." "Yucatec," a Colonial Spanish coinage produced by combining an ending in Nahuatl (the language of the Aztecs) with a misunderstood Maya phrase, was picked. It is a delightfully foolish coinage, for it treats the Maya word *tan* (*t'aan*, "words, language") as if it were the Aztec ending -*tan* or -*tlan*, "place, country." The ending -tek, "person or people," is then added on, just as Aztlan, "heron place," gave us "Aztec," "people from the heron place." Of course, this produces the ridiculous concept of a "Yuca place." There is no such word in proper Nahuatl, nor does "Yucatec" make any sense in Maya.

Moreover, the Spanish form "Yucateco/a" properly refers to anyone living in the Yucatán Peninsula—or, sometimes, to anyone from Yucatán state. The Yucatec Maya should thus be any Maya who live in the Yucatán Peninsula, not just the speakers of *maayaj t'aan*. It is also incorrect to spell the word "Yukatek," as done recently by some scholars. The word is Spanish, and is correctly spelled in Spanish orthography.

But there is no hope of changing the usage now. I use "Yucatec Maya," then, to mean "speakers of *maayaj t'aan*." However, I use "Yucatecans" to mean "people of the Yucatán Peninsula," whatever language they speak. I realize this is somewhat confusing, but at least I have tried to follow existing usage.

Since there are no common or standard English or Spanish names for most of the plants and animals, and many of the farming concepts, discussed in this book, some acquaintance with the Maya language is necessary.

In the sixteenth century, the colonial Spanish worked out an excellent, accurate, and linguistically sophisticated transcription for Maya. It had its drawbacks. In particular, it failed to distinguish long vowels from short vowels, for the very good reason that the Maya themselves rarely do when they speak at ordinary speed. In the mid-twentieth century, linguists and educators began to develop new systems. Finally, representatives of the major Maya languages met in Guatemala in 1988 and agreed on a system that works well for all the twenty-odd languages of the Maya family.

The coming of the new system has produced confusion. Most of the place names in the Yucatán Peninsula were transcribed under the old system. No one is going to rename the towns; Chunhuhub will not become Chuunjujuub, nor will Oxkutzcab become Ooxk'utskaab. Books and journals currently present a mosaic. Some stick with the old system. Some use the new. Some despair, and use any spelling, at random. Some authors try to use the new, but fall back into the old system out of habit, and thus spell the same word two different ways on the same page (even recent textbooks of Maya archaeology sometimes do this).

This book uses the new system, except for established place names.

In the new system, letters are pronounced as in Spanish (e.g., "j" is English "h") EXCEPT:

> b is like English (not Spanish) b, but typically soft or unreleased. (Technically, this is written b', but I follow most writers in leaving off the apostrophe.) At the end of a word, b often degenerates into a glottal stop; *jaaleb* "paca" becomes *jaale'*.

> x represents the sound written in English as *sh*.

> ' indicates glottalization. After a vowel, this means a slight catch in the voice, as in English "unh'unh" meaning "no." After a consonant, it means a sharp check followed by a little explosion. Ch, k, p, t, and ts are commonly glottalized. Glottalized consonants contrast sharply with their unglottalized equivalents; *p'eek*, "dislike," is a totally different word from *peek'*, "dog"—as different as "dog" is from "talk" in English. If you say, with poorly controlled anger, "Get' those C'ATS' out of here," you will have approximated Maya t', k', and ts'. Since the glottal catch is an integral part of the phoneme represented by the preceding letter, the apostrophe has to go before the period at the end of a sentence (copy editors take note!).

In proper Maya, no word can start or end with a vowel; there is, if nothing else, always a glottal catch. This is generally elided in ordinary Chunhuhub speech, especially at the beginning of a word and especially among those who habitually speak Spanish. (Spanish has no phonemic glottal stop and thus no such rule.)

Yucatec Maya is a tonal language, but tones are inconsistent in Chunhuhub, due to migration from various dialect areas. Pending further linguistic research in the area, tones are not marked herein.

One question for the practical writer on Maya plants and animals is whether to indicate the Maya fondness for using the feminine/diminutive prefix *x* (formerly *ix*, as seen in old dictionaries). Maya frequently use this suffix for small, soft plants: *k'anaan* becomes *xk'anaan*; *jeeret'* becomes *xeeret'* (losing the "j"). I find no way to be consistent or rigorous except to leave off the *x* in all cases. Most authors use the *x* when it is common; the problem is that it may be commonly applied in one community and not in another. In Terán and Rasmussen's (1994) community, for instance, the Spanish plant name *lenteja* is usually pronounced *xrenteja*, and they so spell it. In Chunhuhub, however, it is almost always pronounced *lenteja*. I believe the only proper course of action is to be absolutely consistent about leaving off the *x*.

I have endeavored to find the most "correct" pronunciation of each name, rather than resort to the evasion—and it is an evasion—of listing countless variant pronunciations of the same word (as many books do). My "correct" pronunciation is what my friends consistently tell me when they are speaking slowly and carefully so the ethnographer will get it right. Maya are as prone as anyone else to pronounce words in a fast, slurred way, and to use variants, such as the diminutives. In rapid speech, *jeeret'* becomes not only *xeeret'*, but also *xeret'* and even *xeret*. *Jaaleb* becomes *jaleb* or *jale'*. (Cf., in rural English, "hoss" for "horse," "popple" for "poplar," etc.) Remember this when reading what follows, especially if you compare my spellings with those of other authors. I have checked usage against Bricker et al.'s standard dictionary of Hocabá Yucatec, but the dialect of Hocabá sometimes differs from that of Chunhuhub, and my consultants often pronounce words in a nonstandard way; I transcribe what I hear. Bricker et al. had the same problems; *loj*, "ceremony," appears in the dictionary as *loh*, *looh*, and *lo'oh*. (Their dictionary uses "h" for the /h/ phoneme, where the standard system uses "j.") Many other words, too, have the same flocks of variants that they have in Chunhuhub.

As in all my work in local languages, I am reminded, almost daily, of the deathless words of T. E. Lawrence (Lawrence of Arabia): "Arabic names won't go into English, exactly, for their consonants are not the same as ours, and their vowels, like ours, vary from district to district. There are some 'scientific systems' of transliteration, helpful to people who know enough Arabic not to need helping, but a washout for the world. I spell my names anyhow, to show what rot the systems are." (T. E. Lawrence, *Seven Pillars of Wisdom* [New York: Doubleday, Doran, and Co., 1935], 25)

Acknowledgments

A WORK LIKE THIS is such a mammoth collective project that I cannot list more than a small fraction of those worthy of acknowledgment. First, of course, I owe a monumental debt to the people of Chunhuhub, Presidente Juarez, Polyuc, Margaritas, and other Maya towns of Quintana Roo. In particular, I am deeply grateful to my coworkers and coauthors: my landlord and friend, Pastor Valdez, and my field assistants, Felix Medina Tzuc and Aurora Dzib—now both bound to me by ties of *compadrazgo*, the fictive kinship that is taken so seriously in Mayaland. I owe much to my wife, Barbara Anderson, and our children. Among professionals who helped me with all manner of matters in the field, Myra Appel, Salvador Flores Guido, Gene Hunn, and Juan Jimenez-Osornio stand out. Above all, Betty Bernice Faust (of the Center for Advanced Studies, Mérida) was a lifeline and constant support, always ready to compare notes from her communities (especially Pich, Campeche) with my findings. Botanical assistance and identification involved help and support by Gerald Islebe, J. Ortiz, Victor Navarro, Andrés Sánchez, Odilón Sánchez, Juan Tun, and several students (see Anderson 2003). At my home institution, the University of California, Riverside, I had the help of Mayanist and political-ecological colleagues including María Cruz-Torres, Scott Fedick, Alan Fix, Arturo Gomez-Pompa, David Kronenfeld, Karl Taube, and others. Finally, I am deeply indebted to Allyson Carter and Christine Szuter and the editorial staff at the University of Arizona Press for help and support.

I am most grateful to the University of California, Riverside, for funding my research, both through salary (especially during the sabbatical leaves when I carried out the research) and through a long series of small grants.

Special thanks to Scott Atran, who not only helped with constant discussion and sharing of material, but also undertook the heroic labor of reading the incoherent mass of material that accumulated over twelve years of writing, and provided suggestions for turning it into a real book.

The faults and errors remaining in this work are purely my own doing.

Political Ecology in a
Yucatec Maya Community

CHAPTER ONE

The Land of Not Much

A COUPLE OF Leo Tolstoy's stories bear retelling here.

In the first, a king becomes obsessed with learning what is the most important time in his life, the most important person in his life, and the most important thing in his life. After a long series of adventures, he learns that the most important time is *now*; the most important person is *whoever you're with*; the most important thing is *be good to them* (Tolstoy 1962:82–88).

In the second, a small ship is lost at sea. On the boat is a bishop, worrying about his position, wealth, and status. Night comes on, and a storm rises. The sailors see a light and make for it. They find a tiny hermitage on a lonely island. Three hermits live there, continually praying: "Three are we, three are Ye, have mercy on us." They receive the seamen kindly, feeding them generously from their meager store. At dawn the captain finds his bearings, and the ship is ready to sail. The bishop thinks of rewarding the hermits. "Poor silly things, they don't even know how to pray properly. I can at least teach them the Lord's Prayer." This he does, and the hermits are very grateful. The ship sails on.

Night comes again, and the storm wind rises. Suddenly a light appears astern, getting rapidly nearer and brighter. The sailors become afraid. Soon the brilliance resolves itself into the three hermits, wrapped in blazing light, running along the top of the waves with the speed of the wind. They reach the ship, stand in a row, and address the bishop, saying: "Sir, we're sorry, we have forgotten the nice prayer you taught us. Would you say it again, please?" (Tolstoy 1958:468–73).

Tolstoy never saw the Yucatán Peninsula, but the first of these stories captures a great deal of the Yucatec Maya values system. Much of what follows is more easily understood if one recalls that the Maya have long known what the king had to discover.

The second story is a parable for the fate of "development" in Mayaland. Endless files of drop-in outsiders have come to "help" the Maya with schemes that are often genuinely good and valuable, but that are absurdly wrong for the situation. The Maya do not know everything, and are not always as virtuous as Tolstoy's imaginary hermits, but they are most certainly better judges of their needs and competences than a one-night visitor. In Tolstoy's story, the bishop is properly humbled, and goes home with the obvious lesson learned. Rarely, in this imperfect real world of ours, do developers learn so well (Dichter 2003; Frazier 2004).

Outsiders, and some Maya too, have been quick to blame "corruption" or "globalization" or "the United States" for every failure and misfortune in Mayab (Mayaland). The truth is not quite that simple. I have often said that for every bad thing done by bad people being bad, there are a hundred bad things done by good people being stupid. Like the bishop, they often have genuinely good ideas and genuinely worthwhile knowledge. But they do not take the trouble to understand the real situation. Thomas Dichter (2003) has described this problem in world development.

Like everyone else, the Maya of Quintana Roo have had successes and failures in managing their environment. Ancient successes—the rise and maintenance of the Classic Maya civilization—remain to be fully studied, but some information is emerging. Recent successes include some projects that have attracted world attention (notably the Plan Forestal; see chapter 4, also Flachsenberg and Galletti 1998, Galletti 1998). Failures are, fortunately, less notable, but they are instructive (Anderson 2003; Anderson and Medina Tzuc 2004).

The ancient word *Mayab*, the Maya's name for their land, has been given a wryly humorous (and wholly imaginary) folk etymology, tracing it to the Maya phrase *ma'yaab*: "it doesn't have much." In the sixteenth century, Friar Diego de Landa, with his usual combination of sharp insight and acid judgment, called it "all . . . flat stones with very little earth" (Landa

1937 [1566]:93). It took some rather special efforts to build, on this land, the civilization often considered the greatest in Native America (see, e.g., Taube 1992; Houston 1997; Sharer 1994). Agriculture in the region, and in tropical America, has a long history (see Piperno and Pearsall 1998; Sauer 1952). The civilization actually started in better terrain to the south, but its extension and long success in the Yucatán Peninsula still demands explanation.

In the 1980s, the regnant paradigm among Mayanists was that the civilization had flourished on the basis of intensive cultivation techniques that are now lost or almost lost: terracing, raised-bed cultivation, swamp cultivation, soil enrichment of various sorts. There is controversy over how much part these played in Maya success (Gómez-Pompa et al. 2003; Pohl et al. 1996; Sharer 1994, 1996.) What matters most now probably mattered most then: slash-and-burn cultivation, simple in concept but based on a superb knowledge of how to extract every conceivable shred of use-value from a harsh but wondrously diverse and biotically rich environment—a striking combination of sensitivity, awareness, and ability to work hard under rough conditions.

Recent discoveries indicate that urbanization and high population densities appeared in Mayab by two thousand years ago. Classic Maya civilization lasted from 200 to 800 A.D., and then collapsed for complex reasons still not clear (Culbert 1973; Schele and Freidel 1990; Sharer 1994; Webster 2002). Ecological causes for the fall are adduced by many authorities (Santley et al. 1986; Schele and Freidel 1990). Major drought was certainly involved (Curtis et al. 1996; Gill 2000; Haug et al. 2003; Hodell et al. 2001; Islebe et al. 1996). Postclassic cities maintained substantial populations, even in many of the least environmentally promising areas of Maya occupation. Historical documentation of the first observations of the peninsula by the Spanish explorers indicate dense populations in Campeche, Champoton, Tulum, Tiho (Mérida), and other places, including Chunhuhub (Jones 1989; Scholes and Roys 1968; Wagner 1942). Whatever caused the "fall" of Maya civilization, it did not end that civilization; it merely decimated the population of the central areas dominated by it. The civilization not only continued, but it resisted the Conquest;

the last independent Maya state fell in 1697, about 200 years after first contact with Europeans. The Maya kept on farming, in spite of oppression (Stephens 1963 [1843]; Turner 1911), and are farming still.

Maya agriculture and forest management, like all other traditional resource management systems, are well adapted to the needs of the people who practice it. It is not a thoughtless and mindless waste of the environment, but neither is it a modern conservation biologist's utopia (Fedick 1996a). Its purpose is to keep people alive, not to maximize biodiversity; it does maintain high levels of biodiversity, but only because that is the most effective way to maximize the total stream of resources coming from the harsh landscape.

Contemporary Yucatec Maya have their own ways of managing the forest and scrub of the Yucatán Peninsula. They have lost many of the intensive techniques now known to have been practiced by the ancient Maya, such as tree cropping, terracing, and raised fields (Darch 1983; Farrington 1985; Flannery 1982; Kunen et al. 2000; Siemens 1998; Harrison and Turner 1978). On the other hand, they have an extensive knowledge of plants—as food, medicine, and industrial materials—and many types of agriculture, from milpa (see chapter 3 for a fuller explanation of milpa) to home gardens to semi-wild orchards (Faust 2001; Gómez-Pompa 1985; Gómez-Pompa et al. 1987; Hellmuth 1977; Mendieta and del Amo 1981; Pope and Dahlen 1989; Sanabria 1986). The contemporary Lacandon Maya, who may be a group of Yucatec who fled to the deep forests during early Spanish colonial times, also have a strikingly detailed and successful environmental management system, or had until recently (March 1987; McGee 1989; Nations and Nigh 1978, 1980). The spectacular environmental knowledge of the Highland Maya groups of Chiapas is also well known (Berlin, Breedlove, and Raven 1974; Breedlove and Laughlin 1993; Hunn 1977; Lenkersdorf 1996.) Their knowledge represents a vast storehouse of information about ways to use local plants and ways to manage the environment, and much of this knowledge could be useful far beyond the Yucatán Peninsula (Gómez-Pompa et al. 1972). Heirs of a tradition of fierce resistance against outside rule (Bartolomé 1988), the Maya now self-consciously maintain cultural traditions, including those pertaining to land use. The modern transformations and possibilities of these have attracted recent attention, not merely for their local interest, but also because some current techniques and projects are of worldwide significance (Primack et al. 1998).

Traditional Maya religion of the area also encouraged such management strategies. The *Yuntsiloob*—the Lords, including the *Yumilk'aax*, the Lords of the Forest—would punish anyone who selfishly overused resources (Anderson and Medina Tzuc 2004; Faust 1998; Faust et al. 2004).

Maya agriculture is "sustainable" in the sense that it will support a fairly large population indefinitely, without totally ravaging the environment. It will not support an infinite population; indeed, it will not support the current population in many (if not most) parts of the peninsula (see, e.g., Terán and Rasmussen 1994). The exact number of people per square mile who can be sustainably supported is undetermined, but certainly some parts of the Yucatán Peninsula have exceeded the level. Also, Maya agriculture is not "sustainable" in the stricter sense: it does not automatically guarantee that no damage will be done to biodiversity. If one wishes to have not only farms but also mahogany, wildlife, and ecotourism, then further efforts are needed.

However, even the biggest and most vulnerable animals, jaguars and tapirs, still exist. Two Maya friends and I followed the fresh tracks of a jaguar for half a mile once—not a deeply reassuring activity at the time, but reassuring in retrospect, since it shows that jaguars can live with a fairly dense human population.

The cities of Maya Mexico are Spanish speaking, but their culture is not much like that of the rest of Mexico. The Yucatán Peninsula has its own Hispanic culture and traditions; since the Conquest, it has been a region apart (see Roys 1955). Not until the turn of the last century did a railroad link Mérida with the rest of the country. Before that, Yucatán had been more oriented toward Cuba; Havana is closer, and an easier sail, than Veracruz. Habanera rum and habanero chiles are only two of many clues to this Cuban connection. Many Yucatecans still refer to the rest of the country as "Mexico"—in opposition to the peninsula, that is—and use the term "Mexican" as a reference to aliens, not a term of self-referential pride. Sometimes they even use the genuinely unfriendly term *huach* (a widespread marginal-Mexican slur term for central Mexicans; it is derived from *huachinango*, "red snapper fish," a reference to the red coats of nineteenth-century Mexican soldiers).

The Maya and Spanish regional cultures blend into each other in complex ways. Each has its regional variants, and therefore the blending is different from region to region. In northwest Yucatán state lies the desolate and impoverished henequen country, where the blazing white limestone reflects so much heat that the worst sunburns are often on the bottom of the chin. The northeast peninsula is centered around the raw, new, mass-tourism world of Cancun. The south part of the state is a line of ancient, prosperous agricultural towns, tracing a line along the foot of the Puuc—"the hills"—where good soil has washed from limerock cockpit country down to the plain. In the southeast, in central and southern Quintana Roo, lies a land of dense forests in which the last independent Maya held out throughout the nineteenth and early twentieth centuries. Other subregions—Valladolid, Campeche, Tizimin, Chetumal—all have their own intermingled but distinct histories.

Central Quintana Roo is the "Zona Maya." Today, this local name distinguishes it from the "Zona Turística" in the north, and from the Chetumal area in the south, where urbanization and mass migration from elsewhere in Mexico have changed the cultural landscape. The center conserves an agrarian, Maya-speaking way of life, a stronghold of distinctive costumes, ceremonies, and resource management strategies.

Unlike many small tropical towns, Chunhuhub has not been a helpless victim of vast international forces. The people here are descendents of the incredibly heroic bands that stood off the full force of the Spanish conquest (which was incredibly brutal; see Clendinnen 1987; Cortes 1986; Landa 1937; Las Casas 1992; Patch 1993; Restall 1997, 1998). The last independent Maya state fell only in 1697 (Avendaño y Loyola 1987; Jones 1989, 1991; San Buenaventura 1994). Later, the eastern Maya (and some non-Maya) rebelled against the Mexican government in the "Caste War" of 1846–48 (Bricker 1981; Dumond 1998; Farriss 1984; Jones 1989; Reed 1964; Rugeley 1996, 2001). The Maya of what is now Quintana Roo became independent, worshipping "talking crosses" and therefore known as the Cruzob (mixing the Spanish word for "cross" with a shortened form of the Maya plural—*o'ob*). They maintained de facto independence until 1901; parts of the inner peninsula remained independent of outside control into the 1930s. There are, thus, men and women who remember genuine freedom—the days when the Mexican government and its troopers did not dare to go into central Quintana Roo. Today, the towns of the Zona Maya of Quintana Roo have made their peace with the government, but they have done it on their terms. They are prepared to fight again, if need be. Some local militia members still drill here, and in the eastward towns they drill far more seriously—against the day when trouble may break out. Insofar as Chunhuhub is entering the modern world, it is entering with a full sense of self-worth and ability to cope. The Maya feel the inevitable pressures of government and international market, but their response is to fight back—through education, politics, and economic enterprise today; perhaps through force in the future. They remain intensely aware of what their fellow Maya in Chiapas are doing in the present (the Zapatistas are only one of the currents of resistance there). They echo many of the stories told in Paul Sullivan's book *Unfinished Conversations* (1989). Sullivan lived in Señor, a center of resistance even now. Conversations there still turn on battles, betrayals, deceptions, and broken promises. Chunhuhub is farther from any storm's eye than Tusik and Señor are, and our conversations here are less intense and fiery than the ones that Paul Sullivan held there, but our conversations are just as far from finishing. Polyuc's curer, Don José Cauich, who lived in Señor sixty years ago, tells some of the same stories Paul Sullivan recorded. The Conquest, in central Quintana Roo, is not a done deal.

Between the middle 1980s and 2000, Chunhuhub got reliable water, reliable electricity, reliable mail, telephone and television service, and paved streets. Its clinic and pharmacies became good enough to guarantee First World levels of health. More important to the people was the coming of quality education, especially the new technical school (CEBETA). This allowed the education-conscious people of Chunhuhub to enter the new millennium with real opportunities. Modern education has brought many teachers and other educated people to the town. Teachers working in the small forest *ejidos* nearby, as well as teachers in Chunhuhub's own schools, prefer to live in Chunhuhub. Although not all families are positioned to take advantage of the new educational opportunities, all benefit from the new economic activity generated. Nearby, the vast tourist complex spreading from Cancun has disrupted communities, devastated local environments, and brought crime and drugs to a huge swath of northeastern Quintana Roo (Juarez 2002; Pi-Sunyer and Thomas 1997). So far, Chunhuhub, as well as the whole west of the state, has been spared the worst of this. The "zona turistica" influences life in the "zona Maya" primarily by offering new economic opportunities: jobs for educated youth, markets for fresh fruit, new opportunities for selling craft items.

Development of agriculture in Quintana Roo has been notably less successful than development of infrastructure, health, and education. Maize is the main crop, and is raised by methods that go back to the ancient Maya. Attempts to "modernize" this system have met with mixed success. Traditional agriculture spared the land and managed it efficiently, whereas most attempted "modernizations" have not considered this need. The only really successful modern agricultural enterprises have been fruit growing and intensive mixed gardening. The latter has not caught on, largely because many of the people who are intelligent and enterprising enough to do it have gone to CEBETA instead, and have become teachers and technical experts all over south Mexico.

Chunhuhub manages its lands collectively. Members of the ejido—the communal land-managing unit—meet in assemblies to decide policy. At a lower level, families—usually extended families—manage the fields, orchards, gardens, bee and honey operations, forestry and sawmilling, and indeed almost everything in Chunhuhub.

For the Maya, family and community management is an age-old tradition. To this was added the ejido system in the 1930s; in Quintana Roo, it formalized and extended an already-existing pattern of community regulation (see Villa Rojas 1945). Wider-scale government planning has also grown and developed. Planning takes place at all levels. Individuals need family support to get land, work land, and manage forest—and, also, to get an education, start a store, or stay healthy. Other social forms common in Mexico—traditional markets and market relations, informal credit-and-loan societies, and the like—do not flourish in the Yucatec Maya world

(Brown 1988). Only the ejido assembly, the town government, and the extended family system provide solid, dependable social frameworks.

The leading question in economic and social action, for Chunhuhub as for the world, becomes one of trading long-term, wide-flung interests against narrow, short-term ones. There is always pressure to think only of one's immediate nuclear family, right now. Sometimes this pressure becomes overwhelming; if one's children are starving, and there is no other way to get food, one will shoot a deer, even with the knowledge that deer are endangered and protected. One may even cut an undersize mahogany tree, an act that in Chunhuhub is considered to be one of the worst sins against the collective good. Always, one must trade off individual, family, and community interests, and trade off present welfare against future opportunities. Sometimes this is done successfully, to the benefit of all. Sometimes it is not; Chunhuhub cut out its valuable woods in the early 1990s, and suffered for years as a result.

Such decisions are more difficult in a context of rapidly rising population. Quintana Roo's population is increasing 7 percent to 8 percent per year. In such a situation, any plan that allows everyone to take a reasonable share of natural wealth becomes hopelessly inadequate in a very short time.

The general question of narrow, short-term interests versus wide and long-term ones thus becomes the major question facing Chunhuhub, and facing anyone writing about the town. Indeed, it is the major question facing the world today—perhaps especially the parts of the world that still have tropical forests.

Worldwide, mistakes are almost always in the direction of privileging the short-term over the long. At this point, state and federal governments often intervene—sometimes for good, sometimes not. Under the best of circumstances, cooperation between local and national governments can provide excellent "comanagement" (Pinkerton and Weinstein 1995); this has sometimes succeeded in Quintana Roo (Primack et al. 1998), but much more needs to be done.

The decline of tropical landscapes, including the disappearance of the world's tropical forests, is generally agreed to be a serious, even catastrophic, problem of this sort. Whether these forests are managed as common property or as private holdings, they are disappearing with extreme speed. The reasons are complex, ranging from market imperfections to deliberate political choices (Ascher 1999). Mexico has the second highest deforestation rate in the world (after Brazil). Between 1993 and 2000, Mexico lost an area of forest equal to the whole country of Ireland—2.78 million acres ("Mexico Losing Forests Rapidly," *Los Angeles Times*, December 4, 2001, p. A15).

Since most of the peninsula is farmed, and all of it has been in the past, wilderness preservation or protection of "charismatic megafauna" will have relatively little effect on Yucatán realities. The teachings of "deep ecology" and other environmentalist movements oriented toward preserving unmodified "nature" are of little use. They are, at best, irrelevant to the needs of the rural poor; at worst, certain "deep ecologists" regard the rural poor as mere "overpopulation," to be quietly removed from the landscape (see Bookchin 1988; the general program of "deep ecology" calls for population reduction—Naess 1986; clearing rural people from sensitive reserves and sites is a move popular with many biologists). If one sees the rural poor as human beings, not as objects to be eliminated, a different philosophy is needed (Romero and Andrade 2004).

The danger is that the Maya may be sucked into the worldwide rural crisis. This crisis is worse in such countries as India and Guatemala, but it has affected Mexico. It involves community decline, poverty, biodiversity loss, germ plasm loss of cultivated crops, soil erosion, and degradation of water resources. These are different results of a single process. The process itself is, basically, the overuse or misuse of natural resource endowments: soil, water, forests, grasses and herbs, animals, and the other useful and potentially useful items that the environment provides. Failure to cope with exploding population in the tropical and subtropical parts of the world has greatly exacerbated the situation. From the point of view of its probable effect on the future of humanity, the rural environmental crisis appears to be by far the most serious and immediate threat facing our species.

If the rural environmental crisis is not resolved, the world food situation, already precarious (Brown 1995; Brown et al. 1998), will become rapidly worse. Already, about one-fifth of humanity is malnourished, and perhaps half are at risk of serious hunger. Food is not the only commodity at risk. Fresh water and forest resources are directly threatened. Moreover, the danger of floods and similar "natural" disasters is vastly increased. Many of these disasters are not "natural" at all, but the result of decades of land mismanagement. Vast wildfires in Indonesia and Brazil in the summer of 1998, as well as floods in China in 1998 and in Central America in the wake of Hurricane Mitch, finally forced this realization on world awareness (Abramovitz 2001; the point had been made for China long before, e.g., Mallory 1926). They are not "acts of God"; they are acts of humanity.

The most serious dimension of the crisis is overuse. Resources that could be sustainably exploited are treated as wasting assets. This is most obvious in the case of fish and forests. Less well known are the similar over-drafts on soil, brush, grass, medicinal herbs, game, surface and subsurface waters (Gleick 2000), and other vital commodities.

Overuse is due partly to the needs of an expanding population, and, locally, to expanding affluence. However, much of the problem is caused by inefficiency and waste. Such waste is not usually as spectacular as the fires and floods of 1998, but it is the more deadly for being insidious. Imperfect markets lead to elimination of tropical forests, rich in timber and medicinal herbs, for the benefit of low-intensity cattle production (Painter and Durham 1995). Imperfect or undeveloped markets—to say nothing of the prejudices of misguided developers—lead to concentrating production on one or a few bulk commodities, and neglecting possibilities for far more profitable and diversified cultivation systems (Ascher 1999). To this one may add contamination of water by agricultural chemicals and untreated sewage; deforestation of steep slopes, with consequent soil loss and flooding; and countless other problems.

Given that population stabilization is far in the future, and that we must somehow feed a rapidly increasing number of mouths, there is no hope of going back to a simpler, less productive food system. The dreams of ending globalism and returning to national self-sufficiency have long ago become futile. Most nations do not, and cannot, support their current populations from their own resources. Nor is there any real hope of increasing world resource production much beyond current levels. Existing resources are being rapidly reduced, not expanded. *The only hope lies in greatly increasing the efficiency of resource use.* We must get much more value out of each log, each gallon of water, each square meter of soil, and each diverse local ecosystem.[1] This requires attending to the traditional systems that have made efficient use of local resources. Such systems have also saved biodiversity, and are being put to work, worldwide, for saving it today (Stevens 1997); this is an important part of management today in the Yucatán, with Maya strategies being drawn on in biosphere reserves. Such systems are also labor intensive, eliminating the "jobs versus environment" problem (Goodstein 1999).

Rural environmental decline has been blamed on "colonialism," "capitalism," and other such factors. Unfortunately, it is more pervasive than that.[2] Destructive practices have been invoked by communists, capitalists, and feudalists. Sometimes this was because they were thought to raise production—or, more accurately, to raise production in the short term for a few highly desired commodities, even if this led to decline or destruction of other commodities.

For better or worse, highly directive governments and international agencies are the order of the day, and there seems to be no immediate way of changing this. Moves toward a so-called free market (as advocated by the World Bank, the International Monetary Fund, and similar agencies) have always, in practice, turned out to be moves toward an international order ruled and determined by the agencies themselves and by the rich nations and giant multinational firms that hold so much of the power in these agencies as in the world market. A market dominated by a few giant government-backed firms is not a "free market"; the "new world order" is mercantilist, not "free," and the new international order has its problems (Hancock 1991; Stiglitz 2003.)

Biotic diversity, an outstanding feature of tropical forests, is typically lowered as the forests are progressively transformed. Yet traditional subsistence agriculture may maintain substantial diversity in at least some ecotypes (see Nabhan et al. 1982 for the general case; for the Maya, Rico-Gray and Palacios Rios 1992). The relationship of diversity to ecosystem stability has been the focus of considerable theoretical argument and remains controversial (see, e.g., Altieri 1980; Murdoch 1975). Maya subsistence agriculture is associated with considerable diversity both of cultivated and of wild plants. Maya I know recognize the need to maintain diversity (see also Faust 1998). Given the seriousness of the crisis in tropical forest environments, we must seek out adjustments that have been successful in maintaining a high density of settlement over time (Klee 1980; Laird 2002; Marten 1986; McCay and Acheson 1987; Wilken 1987; Williams and Hunn 1982).

While the Yucatán Peninsula's biodiversity is not in a class with that of the Upper Amazon, or even with southwest Mexico, the peninsula is extremely rich in plant and animal species. It has, for instance, more than a dozen endemic or near-endemic birds, and several other bird species found nowhere else in Mexico. Its forests blend into those of Guatemala, Belize, and far southern Mexico, creating by far the largest chunk of tropical forest surviving in North America—an area acutely endangered, and targeted by preservation efforts around the world. Vast tracts in all three nations have been set aside for protection of various kinds, but the level of enforcement varies. This forest is not only a conservationist's treasure house; it is also a treasure house of mahogany and other precious woods, of genetic resources (from wild allspice trees to tough local varieties of corn and beans), of medicinal herbs, and of countless other useful commodities.

Thus, to the needs of agriculture, planners must add concerns over logging, ecotourism, biodiversity protection, medicinal plant harvesting, and archaeology, among other matters (Primack et al. 1998). Moreover, none of the stakeholder groups speaks with a single voice. One hears acrimonious verbal fights between proponents of rival conservation schemes. Ecotourism, simple in concept, is not simple in execution—does one accommodate vast hordes with a slight interest in the environment, or small groups of dedicated naturalists, or both?

Government officials in Mexico usually come from the middle- and upper-class Spanish-speaking sector of the populace. They often have little understanding of indigenous subsistence cultivators, and still less of the virtues of traditional agricultural systems. Frank anti-indigenous bias is rare in Quintana Roo (though subtle bias exists), but Mexico is a highly centralized country. Decisions tend to be made at the national level or, at the very least, at the highest levels of state government. The national government in particular has historically been less than sympathetic to the interests of local indigenous cultivators. Quintana Roo, to its credit, has been more aware of the strengths of Maya farming systems. This is not unrelated to the active participation of the Maya in political life. But even in Quintana Roo, development until very recently was dominated by external and highly inappropriate models. Lying behind these is the belief that anything devised by urban-educated Spanish-speaking technocrats and bureaucrats must be better than anything done by indigenous *campesinos* (Gonzalez 2001). Fortunately, this attitude has greatly changed in recent years.

Recently, considerable attention has been paid to traditional ecological knowledge and traditional resource management techniques. Unfortunately, all too often, traditional management of biotic resources has been stereotyped as "harmony with nature" or as wanton destruction.

These extreme positions are easily dismissed, certainly in the Maya case. The position that "primitive" people are "in harmony with nature" is now not held, to my knowledge, by any scholars working in this research area. It survives among some New Age devotees and deep ecologists, but it is rapidly dying out even in the more romantic media.

Not long ago, however, cultural ecologists held views that tended in this direction. Highly functionalist interpretations of traditional resource management were general in the literature. Julian Steward, the founder of cultural ecology, had a more dynamic view of traditional ecologies; he saw change, political negotiation, and systemic imbalance as the norm (Steward 1955). Steward's followers, however, tended toward an approach strongly influenced by structural-functional theories that were widespread in social anthropology at the time (see Harris 1968 for examples and chronicles). These theories held that traditional institutions served to keep the social system stable and smoothly functioning (Rappaport 1984).

Cultural ecologists were struck by the similarities between these and the theories of harmonious, stable ecosystems that dominated ecology at the time. Biologists postulated complex feedback loops that kept energy flowing smoothly and community structure stable within natural ecosystems. The key word of the day was "homeostasis." Cultural ecologists argued that the same devices—social and biological—kept the human-influenced ecosystem homeostatic as well (Flannery 1972; Rappaport 1971). Many human institutions were "explained" (one must now add the scare quotes) as devices that served to keep human-environment systems stable, often through negative feedback (particularly famous examples are found in Harris 1966, 1968).

In particular, religion was seen as a storehouse of ecological wisdom, and as a system whose rituals and moral teachings often supported conservation. This view was broadly shared with many historians and students of religion, who saw traditional religions as embodiments of ecological wisdom (e.g., Hughes 1982; Tucker and Grim 1994).

The classic statement—with supporting ethnography—was Roy Rappaport's book *Pigs for the Ancestors* (1984), which, while far more temperate than the New Age excesses, did maintain that the Tsembaga Maring of New Guinea had developed a system that worked smoothly, harmoniously, and nondestructively, through religious construction—and all without the Tsembaga being fully aware of the reasons. This book originally appeared in 1968; the second edition in 1984 added a long and detailed afterword that evaluated criticisms of the functionalist model. Rappaport, while admitting that he had been perhaps too Panglossian and simplistic in places, defended his view. A final, brilliant, complex statement of Rappaport's position appeared posthumously in his work *Ritual and Religion in the Making of Humanity* (1999), but its message had gone far beyond the narrow confines of ecological anthropology.

"To every action there is an equal and opposite reaction," and in this case it was not long in coming. A new generation of cultural ecologists, as well as biologists, evolutionary ecologists, and other students, found disturbing evidence of degradation in traditionally managed ecosystems—not least in the Maya case. They also found that many traditional small-scale societies simply do not have conservationist ideas, or have rudimentary ones (Alvard 1995; Alvard et al. 1997; Headland 1997; Kay and Simmons 2001; and, for an extremely exaggerated case that dismisses all Maya wisdom, see Redman 1999). Admittedly, reported societies of this type are small and mobile, and live in areas difficult to manage. Most of them occupy the more remote parts of the Amazon rain forest. Others have been drastically affected by outside pressures, and have probably lost much of their ideological equipment. However, they exist, and they certainly disprove any naive notion of the universality of conservation in traditional societies.

Moreover, modern conservation biologists imposed new and much more stringent requirements for human conduct in ecosystems (Kay and Simmons 2001). Rappaport and his colleagues had concerned themselves with whether the traditional agricultural system could keep producing goods and feeding the people who invoked it. Modern ecologists and conservation biologists are often preservationists. They censure traditional people for any and all deforestation, landscape change, or inadvertent damage to animal populations. This automatically places all agriculture, and most hunting, on the negative side of the balance.

Meanwhile, social and ecological theory turned away from functionalism. Scholars such as Jon Elster (1983) and Jonathan Turner and Alexandra Maryanski (1979) sharply critiqued functionalist social theory. Among other things, they pointed out that one searches in vain for the happy, harmonious, frictionless societies implied by the model. Meanwhile, ecologists had equal difficulties finding stable, smoothly functioning ecosystems. In nature, disruption is the norm, and many an ecosystem is the way it is only because of frequent and unpredictable disturbances, such as wildfires and hurricanes (Botkin 1990). Certainly, the Quintana Roo forest has had to respond to both those disrupters for countless millennia. To that degree, it was preadapted for human disturbance, and humans mimicked nature in their local clearing and burning.

Social theory swung toward the opposite extreme, rediscovering individual rational choice theories, or emphasizing the conflicts that rampant individualism produces. Mancur Olson (1965) held that social groups dedicated to causes were virtually inconceivable, because everyone would free-ride unless given substantial "side benefits." He thus predicted the imminent demise of such organizations as labor unions and the feminist movement. This view in turn has had to be qualified, to say the least (Frank 1988; Green and Shapiro 1994), but it was adopted by many cultural ecologists, who mistakenly saw it as somehow an entailed corollary of Darwinian theory (Wilson 1998). Ecological theory, too, tended toward a more individual-oriented, conflict-aware viewpoint, with competition rather than homeostasis as a central concept. Extremists returned to Herbert Spencer's "nature red in tooth and claw" and "survival of the fittest" images.

Such simplistic views are as easy to critique as the simplistic static-functionalism they replaced. Obviously, society exists; Olson was wrong. Ecosystems do exist; they are dynamic and fast changing rather than stable, but mutual dependence and mutual accommodation often characterize them, just as competition does. (The fact that the mutual dependence must have evolved through selection on individual species does not make competition somehow more "basic." Inclusive fitness is the only thing that is "basic" here.) Neither society nor ecosystems functions with perfect, frictionless smoothness.

One could argue that the more extreme views were never intended to be taken as literal, final truth. They were models that were created to draw attention to some part of the social or ecological process. This is where politics enters in. During the 1950s and 1960s, harmonious function was a goal and a dream, and models that stressed it became popular. Humans were seen as naturally desiring peace, love, and flowers. War was decreed by sadly miseducated leaders; "wars will cease when men refuse to fight," said a common bumper sticker. In that idealistic era, ecology was a science of harmony and mutual interaction.

During the bitter post-Reagan, post-Thatcher years, disillusion and cynicism set in. Scholars of that generation sought models that stressed the innate evil of human nature. To these scholars, it was inexorable biological law—not just the current political fad—that doomed us to savage, cut-throat competition and self-destructive selfishness. A book title, *Demonic Males* (Wrangham and Peterson 1996), captured the spirit of the age perfectly. The psychological mechanisms behind such swings in intellectual fashion are easy to see in retrospect.

In the post-Reagan climate, scholarly critiques of the "traditional harmony with nature" views produced intemperate rhetoric in certain quarters, and, inevitably, in the popular press. Some of the attacks on traditional people as managers can only be described as crude racism. So were many of the "noble savage" writings, as Charles Kay (2001) has pointed out. Simplistic stereotypes, whatever their content, are insulting. On the negative side, the zoologist Kent Redford launched a scathing attack on Native Americans, and accused those who saw them as able to manage resources as propagating the "myth" of "the ecologically noble savage" (Redford

1990—an article wisely and effectively refuted by Lopez 1992). To his credit, Redford has since learned better and greatly tempered his views (see Redford and Mansour 1996). Unfortunately, Redford's sarcastic phrase has caught on. Other attacks were simply silly (e.g., Ridley 1996, who ridicules the idea that primitive people ever manage anything, and then in the very next chapter gives several examples of their doing exactly that).

Implicit in all this literature was the old, long-discredited view that traditional people have been changeless since time began. The idea that traditional people could learn from their mistakes was foreign to both the "noble savage" and "ignoble savage" camps. Thus, throughout the book *Wilderness and Political Ecology* (Kay and Simmons 2001), authors address the extinction of the Pleistocene megafauna as directly relevant to—indeed, part and parcel with—contemporary Native American hunting practices. Evidently, the authors assume that nothing of significance changed in the twelve thousand years separating the two. In fact, there were a few changes that might be considered relevant, such as the invention of agriculture and the subsequent rise of population by several orders of magnitude, but these go unmentioned in the book. If people learn and change, they are hard to stereotype!

The popular media, always ready to give enthusiastic circulation to any negative "findings" about nonwhite, traditional, and indigenous peoples, have given enormous and exaggerated play to such ideas. Never mind that the Maya have maintained high population densities for five thousand years in one of the most unprepossessing habitats on earth, and this without exterminating a single species; the fact (and, to be sure, it is a fact) that they mismanaged the Copan Valley in the Late Classic period "proves" that traditional people (all of them?) are just as bad as the rest of us (see, e.g., Redman 1999 for an outrageously exaggerated claim to this effect, refuted by Gill 2000).

On a more serious level, this was also a time when many conservation biologists were reacting strongly against the tendency of governments and nongovernmental conservation organizations to entrust fragile lands to their indigenous stewards. The conservation biologists feared—not without reason—that this was a dangerous move. Giving a blank check to anyone, even the most apparently conservationist of groups, seemed dangerous. Quite apart from the case of groups with less than conservationist traditions, there were too many cases in which a traditional group with the best conservationist values produced a younger generation that was quick to discard tradition and pick up urban ways, including short-sighted attitudes toward the environment. This led to conflict within the ranks of conservation biologists, between those who continued to respect indigenous wisdom (e.g., Gómez-Pompa and Kaus 1990) and those who did not (e.g., Terborgh 1999). It then led to a subsequent, and not always minor, conflict between those who support indigenous rights no matter what the damage may be if traditions change, and those who recognize that "indigenous" credentials cannot excuse irresponsible and destructive behavior (especially when it represents an explicit break with tradition).

Trying to sort out any semblance of truth from these conflicting theories is not easy. However, fortunately, many scholars continued to carry out close and detailed ethnographic studies. These studies have shown that many traditional and indigenous groups are excellent and self-conscious managers (see, e.g., Berkes 1999; Colding and Folke 2001; Lansing 1991; Lentz 2000; Nelson 1983; Ostrom 1990; Reichel-Dolmatoff 1967). Colding and Folke (2001) review a vast number of rules from a vast number of groups; these rules are conservationist to varying degrees. None is optimal by modern standards, but none is irrelevant, and many (if not all) were evidently effective. The case is certainly made that traditional peoples manage and save. But they do it pragmatically, flexibly, and usually less than perfectly. And they range from superb (though never perfect) managers to wanton destroyers. Even in the same group, one can find savers and wasters.

In religious studies, as well, compilations of evidence from around the world show that ecological teachings are widespread in world religions (Callicott 1994; Kinsley 1995), but do not find a religion that lays out all of modern conservation biology. Perhaps the closest case is Bali, where irrigation is managed by temples. The accumulated agricultural and hydrological wisdom of the Balinese is stored in temple records and, above all, in the learning of the priesthood. This is not unconnected with Bali's conflict-ridden history; only the temples were above secular conflicts, and thus only they had standing to regulate water use. Stephen Lansing's remarkable study of this case (Lansing 1991) is one of the best studies of traditional resource management in action, and it shows that traditional managers do very well indeed. In fact, outside attempts to "improve" the system had to be abandoned because the traditional ways were so much better. (This is not to say that the traditional ways were perfect. Improvements were eventually made, and more are proposed.)

Particularly instructive were the findings of Patrick Kirch and other students of the prehistory of Polynesia (Kirch 1994, 1997). They found, on island after island, a pattern of initial lavish use of resources; subsequent overdraft or extinction of vulnerable resource species, leading to human population collapse and poverty; and, following that, a long, painful uphill climb toward fine-tuned, stable, sustainable systems for using island resources, often involving a large component of imported crops (the local valuable resources having been depleted). This is one of the clearest of many studies that show a strong learning curve in traditional resource management. Moreover, Polynesia shows a variety of responses; each island group had a different history, in spite of cultural similarities. In Polynesia, as elsewhere, change sometimes caught people flat-footed, though more often they learned from experience and went on to better management strategies. Sometimes they practiced heroic restraint and self-sacrifice; sometimes they succumbed to the most trivial temptations, killing off bird species for their plumes.

Traditional and indigenous people are people, like everyone else. They are neither devils nor saints. They try to do right, but succumb to temptation when the lure is too strong. Traditional cultures have developed more or less stable, sustainable ways to support their bearers, or they would not be "traditional." A self-destructive way of life cannot become a "tradition"! However, a traditional land-use system may be terribly destructive to large game animals and old-growth forests in the neighborhood, if it does not value them or depend on them.

Those who work directly with traditional societies (including the present writer) often take, without always making it explicit, a "wise use" view. We look at how the people in question manage to keep themselves going without totally wrecking their environments. Many of the critics are more interested in the animals and plants or the environment as a whole. They see humans as disturbers of nature, not as creators of artificial but interesting ecologies. Many scholars hold nuanced views somewhere between these extremes. Recently, a small, compact group of geographers set out to review Native American landscapes in pre-Columbian America. Though they worked together, producing a three-volume set that is remarkably consistent in approach, and though they all had nuanced and reasonable views, they came to rather different conclusions. William Denevan, reviewing South America, took a "wise use" view and came out broadly positive about Native American management (Denevan 2001); Thomas Whitmore and B. L. Turner (2001), reviewing Mesoamerica, took a view that looked at the whole environment (not just its use to people) and accepted many highly critical studies of Maya and other Mesoamerican systems, and came to quite negative conclusions; William Doolittle (2000) in North America maintained a balanced, neutral, minimally judgmental tone, but by implication sided with the "wise use" view, since he documented enormous change and modification without resorting to the highly negative opinions that surface in such comparable works as Krech (1999) and Kay and Simmons (2001). One's vantage point determines one's views, to some extent.

A prerequisite for effective management is an expectation that one's actions will result in the intended consequences and that the perceived costs of the management strategy to the actor will be justified by perceived benefits. Decision making can be studied as the behavior of "rational individuals" (when individuals perceive material cost-benefit outcomes directly favorable to them), or as an institutionally constrained and institutionally guided process (Barlett 1982; North 1991; Randall 1977). In this latter approach, not only local government, but also local ideology, function as institutions. Behavior can be sanctioned by nonmaterial "goods" such as prestige or emotional satisfactions.

Whether individually or collectively managed, the land and its natural fauna and flora constitute, in central Quintana Roo, a common resource. Study of common property management has advanced greatly in the last several years (Keohane, McGinnis, and Ostrom 1992; Ostrom 1990). In addition to the development of abstract models (outside the scope of this volume), there are many studies of successes and failures of management among local communities (Burger et al. 2001; Ostrom 1990).

According to some theories, collective management is doomed to fail. This conclusion follows from Garrett Hardin's famous "Tragedy of the Commons" (Hardin 1968). Thus, the world's fisheries are depleted by overfishing, even though the problem (too much catching power) and the solution (limit fishing somehow) have both been well recognized for decades (McEvoy 1986; Safina 1997). It has been pointed out, however, that this is a problem of uncontrolled, unregulated access—in other words, of failure to manage—not of collective management. Hardin has corrected his earlier position in an article significantly titled "The Tragedy of the *Unmanaged* Commons" (Hardin 1991; his italics). Elinor Ostrom and a number of her associates and coworkers have made the same general point in more detail (Burger et al. 2001; Ostrom 1990). In fact, in most cases, the tragedy of the commons is actually a tragedy of governments and communities refusing to take resource management seriously (Ascher 1999; Ostrom 1990). Study of regulation of commons, in the manner of the contributions in McCay and Acheson (1987), the Conference on Common Property Resource Management (1986), and Evelyn Pinkerton's studied of comanagement (1989), has begun for the Yucatec areas (Atran et al. 1999; Faust 1998).

Bobbi Low has developed a theory that predicts when small-scale traditional societies will attempt to conserve and when they will not (Low 1993). Very simply put: if a resource is so abundant that people see no problem with its future availability, it will not be conserved. For example, conservation seems lacking among the Machiguenga (Allen Johnson, mss. and personal communication 1983; Terborgh 1999) and Bora (Janis Alcorn, personal communication 1990) of South America. Alternatively, if one's conservation efforts are likely to be exploited by outsiders or cheaters, the costs of conservation are likely to be avoided, producing a "tragedy of the commons." Neither of these scenarios fits the case of Chunhuhub. Its resources are limited, its citizens are quite aware of this, and they have enough control over their land to prevent the "tragedy of the commons" if they so desire (Anderson 1992a, 1992b).

This last issue is highly relevant for groups like the Maya, who have been involved with state-level societies for two thousand years or more, and thus are subject to regular expropriation from outside their small rural communities.

This brings us back to Rappaport's insights about the role of institutions, including religious ones. The Huastec Maya, for instance, have a strong ideology of managing resources for wise use and sustained yield, and maintain it in the face of modern change (Alcorn 1984). A full description of such resource management systems thus goes beyond individual decision making to look at various cultural or group representations (cf. Williams and Hunn 1982). These may range from classification and knowledge of biota (for Maya groups see Alcorn 1984; Atran et al. 1999; Berlin, Breedlove, and Raven 1974; Hunn 1977) to settlement patterns, land use, and farm geography (e.g., Carter 1969; McBryde 1947; Wilk 1981). Information flow within systems has been studied (Alcorn 1984 and Wilken 1987 cover some Maya areas), but there is little on such key issues as education (Ruddle and Chesterfield 1977).

Kay Milton (2002) has shown that people have to be emotionally involved in the environment to manage it well. Her arguments are detailed, complex, and brilliant. She discusses in thoroughgoing detail the mix of emotion and cognition that is human thought. She points out that any politically charged question stirs up emotions. She deals at length with the ways people come to love and fight for particular environmental amenities. Her findings mesh perfectly with those of Berkes, Lansing, and others. Religion serves to construct collective representations, to involve people intensely in their social orders (Durkheim 1995). And it is a "total institution" in many, if not most, traditional societies. Thus, it is naturally the institution put to use in saving trees or animals for the future. It and other emotionally compelling institutions are used to enforce long-term, wide-flung considerations against short-term, narrow ones (Anderson 1996a). In the contemporary world of which Kay Milton writes, political ideology has largely replaced religion as the way to get from love and knowledge to social rules. Perhaps that is inevitable. Be that as it may, religion continues to serve among the Maya, Zapotec (Gonzalez 2001), and other indigenous groups. Time will tell whether political ideology and the "rule of law" can replace it effectively.

Perhaps it is time to return to Roy Rappaport's vision—qualified perhaps by the findings of the last couple of decades. As all who knew the late "Skip" Rappaport will agree, he was nothing if not an idealist. His vision of human communities integrated into the world ecosystem was part of a hope for a world more harmonious, decent, and civil than this one. He was saddened by the cynical turn in anthropology and biology, which, I believe, he saw as a tacit admission by many anthropological and ecological thinkers that they had no more nerve to stand against a popular ideology of cruelty and selfishness. Alas, there is perhaps more cruelty and selfishness in the world than he let himself admit. But he was right in saying that there are alternatives, and that human societies have not been totally unable to see or find them.

The question posed by recent debates is the wrong one. No one should ever have wasted precious time debating whether traditional indigenous people are "good" or "bad" for the environment in some abstract sense. The questions must be: what can we learn from these peoples, who have, after all, been living on and with the land for thousands of years? What do they do that works well, and where do they fail (teaching us what we should we avoid)? And how can we work with these peoples in the places where they still manage their traditional lands (Redford and Mansour 1996)?

It is safe to say that every human group, every culture, has its share of unique and valuable environmental knowledge. The investigations of ethnographers and ethnobiologists into traditional ecological knowledge have been sadly few and inadequate, but they most certainly have established the richness and value of indigenous traditions.

Moreover, most groups do have valuable ideas about ways of managing and conserving resources. In too many cases, this knowledge is rapidly dying out. Shotguns, the cash economy, and similar innovations make overexploitation easier. Rapacious governments, racist paramilitary units, and sheer illegal land grabs dispossess or impoverish indigenous groups, destroying their knowledge in the process—often by simply killing them off.

Ideological influences, from narrowly interpreted Christianity to Hollywood's and Disneyland's plastic-wrapped dreams, replace the indigenous worldviews. The result is that we are losing, every day, a large percentage of humanity's accumulated wisdom (Plotkin 1993). In this "Information Age," computers transmit and store trillions of bits of information about movie stars, sports, business transactions, and fads, while humanity's vast store of practical and valuable knowledge erodes rapidly. Within a generation, most of it will be gone.

The scholars who viewed traditional peoples as reservoirs of ecological perfection were at least directing attention to the value of studying cultural traditions. The new overreaction has irreparably, perhaps fatally, damaged the quest for traditional ecological knowledge. Too many scholars and change agents are assuming that there is nothing worth seeking. Even more tragic is the damage to the attempts to utilize traditional ecological knowledge in formulating better plans to manage the environment.

Management is a moral issue. It is about deferring gratification. It is about planting trees now in the hopes of having more later; about sacrificing the immediate interests of the few to the wider interests of the many; about determining where your rights stop and mine begin. It is, in short, the old question of trading off short-term, narrow benefits and long-term, wide-flung ones. This cannot be a wholly rational choice, in the economic sense of "rational"—best means to predetermined ends. The ends themselves are part of the question. Someone must decide whose interest is to be considered, whose to be protected. Someone must decide what to do when one person's resource use hurts another.

The most difficult of such questions is what to do when current users cry "foul" over a conservation scheme that cuts back their take. If they are sport hunters, they usually get little sympathy. But if they are subsistence farmers who will be severely impacted by legislation to protect forests or game, the case is different. When individuals' livelihoods are at stake, the benefits of the conservation scheme in question must be extremely clear and extremely sure. Schemes that "might possibly" pay off in the future are difficult to justify. One cannot morally ask people to make huge and concrete sacrifices because of a gamble.

On the other hand, overgenerosity to current users, while understandable, is suicide in the long run. Mexico has consistently felt that existing users whose livelihood depends on a resource should get first consideration over conservation. This is understandable, even reasonable, given Mexico's severe problems with following through on conservation and development plans. Unfortunately, the attitude has been pushed to extremes. Mexico has good hunting and game management laws, for instance, but they remain on paper (Alvarez del Toro 1991); they are never enforced, and the Maya (including law enforcement agents) I interviewed do not even know what they are.

All these ecological problems take place in a context of rapid globalization. The media often promote a view that globalization is a new and unprecedented phenomenon. However, world-system economics is no stranger to the Maya, who ran their own "world system" with stunning success in long-vanished centuries. By 600 A.D., Maya trade and contact extended from central Mexico far down Central America. Later, Columbus met large Maya canoes plying Caribbean waters with cargoes of many sorts. The region's colonial period brought contact with the Old World. Finally, the Yucatán Peninsula became completely integrated into the modern world system as long ago as the mid-nineteenth century, when henequen rose to prominence in international trade.

Yucatec migrants work in Los Angeles, Yucatán honey sells well in Germany, and Quintana Roo mahogany appears in guitars in New York. Decisions about henequen in Brazil and Tanzania continue to impact what little is left of the Yucatán's most famous crop. Maquiladoras, tourist hotels, and chain hamburger stands spring up like weeds in a fallowed field.

The Maya share southeast Mexico with a large and diverse population of Spanish speakers, to say nothing of immigrants from other ethnic backgrounds. There are also the vast hordes of tourists that sometimes, during midday hours, outnumber the Maya in their own villages (at least at Chichen Itzá and Cobá). Belize and Guatemala are close neighbors, increasingly important as more and more decisions are internationally negotiated. Meanwhile, many young Maya migrate to the United States and even farther afield. A rising Maya elite includes European-educated intellectuals. All but the humblest Maya households have television. Washington politics, European soccer, and Mexican movie stars are more common topics of conversation than the Yuntsiloob, *aluxoob*, or other Maya spirit-beings.

The Maya find themselves at the beginning of a new millennium—not only in Western terms but in their own, for the current Great Cycle of the Maya calendar comes to an end on December 21, 2012 (Sharer 1994:568). This cycle began with the creation of the world as the Maya know it, in 3114 B.C., a date interestingly close to the time when agriculture suddenly became important in the Yucatán and south Mexico.

From all this postmodern confusion of tongues, icons, and dreams, a new world must be made—shaped from whatever arises with the dawning sun on that fateful postsolstice day in 2012. Current trends in Mexican politics and world environment hold much that is troubling. The Maya will meet the challenge with the same courage, toughness, and good nature in the face of adversity that has carried them through the last Great Cycle's wild swings of fortune.

CHAPTER TWO

Chunhuhub
The Environment

CHUNHUHUB lies on the break point between the low hills of the interior—continuous with the archaeologically famous Puuc of southern Yucatán—and the flat plain of coastal Quintana Roo. The western part of the ejido reveals a classic karst landscape of steep rounded hills and small valleys and sinkholes; relief is about 100 meters. The eastern part appears from any distance to be absolutely flat. On the ground, one discovers a very small-scale but hard-to-traverse karstic landscape—relief is a very few meters at most, and is jagged and rough.

Chunhuhub is one of a long string of towns that curves around the foot of the Puuc hills. Though it is 90 km from the sea, the town is only 20 m above sea level. Above it, rough limestone domes form the core of the peninsula. Upthrust along ancient faults, they rise only some 50 m in Chunhuhub, rising to 118 m slightly to the west. (The highest point in Mexico's Yucatán, some 365 m, lies far to the southwest.) Still, they look impressive enough, rising suddenly from the vast flat limestone shelf that extends east and north to the Caribbean.

Chunhuhub's climate is tropical, but seasonal. Winter temperatures fall to an average of 22.7°C in December and January. The hottest month is May, when the cloudless skies of the searing dry season bring the average to 28°C (82°F). The year-round average is 25.8°C (78.5°F). Daytime variance ranges from virtually nil in the height of the rainy season to 10° or even 15° on dry winter days when nighttime temperatures fall rapidly. The most dramatic changes occur when winter winds shift from south to north; a strong norther brings cloudy skies and temperatures in the teens

(Celsius; around 50°F), which can feel really cold to people sleeping in hammocks and unused to chill. (Hammocks, being uninsulated on the bottom, are very much cooler than beds.)

Chunhuhub has all four seasons, well defined. Spring begins warm in March and rapidly becomes hotter, climaxing in the searing dry seasons—*yaaxk'iin* ("first season") in Maya, *seca* in Spanish—of April and May. The rains begin about mid-May and are reliable and heavy by mid-June, continuing to a peak in September, when hurricanes may bring enormous floods (as in 1989 and 1995). (This makes September the rainiest month on average, with 20.31 cm, but July and August are actually more reliably rainy.) Fall is a brief time of transition. Winter, essentially November to March, is actually the driest season in terms of rainfall (the low is 2.58 cm in February), but feels moist because of weak sun and the frequent cloud and drizzle brought by northers, and sometimes by winds from other directions.

Wash off the hills has concentrated alluvium in the low areas at their feet. Here the soil is extremely fertile, and often deep and rich in organic matter. Drainage is excellent. Such lands have been magnets for settlement for thousands of years. In Yucatán state, they are the mainstay of the line of old towns—Ticul, Oxkutzcab, Tekax, and lesser settlements—that are the centers of fruit and vegetable cultivation. In Quintana Roo, their potential is yet to be exploited fully.

Rainwater is acid enough (from absorbing CO_2 from the air) to dissolve limestone. Chunhuhub has plenty of rain—111.6 cm (44"; I think this is an undercount) a year—but the rain immediately seeps into myriad cracks and holes, rapidly reaching the water table, which itself is barely above sea level. Water has eroded passages and tunnels to the sea, and these naturally have sought the lowest level. Chunhuhub's water table is more than 100 meters down, around sea level.

There is no surface water of any kind. The nearest *aguadas*—water holes—lie in Dos Aguadas, the neighboring ejido on the west. The nearest *cenote* (large limestone sinkhole) is in Polyuc to the north. No streams, not even transient ones, exist anywhere in the area. Water sinks into the honeycombed limestone so fast that rainpools formed by torrential rains are absolutely dry in a few hours. Even the hollows in rocks in the forest do not supply water for long after rains. Persons out in the forest, when short of water, rely on *saya ak'*: wild grapevines that store water in their thick stems. The area is thus absolutely uninhabitable to humans without tanks or wells.

All the many Classic communities in the area—including Chunhuhub itself—reveal large reservoirs. Many probably possessed drainage-catchment systems like those of such sites as Palenque in Chiapas. The ancient Maya coped with the water problem by settling in fairly high, well-drained sites that were next to an aguada if possible; failing that, they dug a reservoir. They developed the aguadas to improve their water-storing capacity, as they did elsewhere. The huge Nueva Loria archaeological site is next to a large lake, now prone to dry up or at best go stagnant. One assumes that it was deepened and managed carefully in the past; otherwise it could not have sustained such a site. Other aguadas I have seen have stone-lined banks and dug-out, leveled bottoms.

In this modern world, people dig wells. Lucky people hit a perched water table, which not only means a short way to water (as little as 10 meters) but also clean, pure water. Such perched water tables develop where clay lenses in the limestone trap fresh water. Less lucky people have to go down some 50 to 100 meters, to more usual groundwater levels, and the water there is polluted. The water is hard enough to leave a white deposit of lime in a pot used only once to heat a few cupfuls.

In the 1980s, Chunhuhub got its pumping station and piped water. This saves people from hauling buckets up 50 meters, or trying to get by on rainwater and thus living a dirty and uncomfortable life. People still save rainwater; it is more pleasant for bathing. Also, it cooks beans tender. Hard water means hard beans. The pumping station drains underground aquifers about 50 meters below the surface. In spite of this depth, the water is thoroughly contaminated, especially with amoebas and *Giardia*.

When Chunhuhub does have surface water, it often has too much. The torrential rains of Hurricane Roxane, in September of 1995, inundated all the fertile hill-foot valleys. Some of these turned into rivers draining the hills. Crops sustained losses of 50 percent to 100 percent. Water persisted in low clayey places for months afterward.

The lack of water in Chunhuhub is in striking contrast to the situation slightly to the south. There, a wide band of low clay-floored terrain cuts across the peninsula, producing a continuous chain of lakes, marshes, and wet savannahs. In the early colonial period, the place where the road crossed this zone was known as the "horse bone area" because so many horses got stuck and died in the mire (Jones 1989:137). At present it is used for extensive cattle ranching and some rice farming. Several Chunhuhub families have cattle ranches in this area.

The Maya are well known for their extensive and detailed soil classification system. Maya cultivators spend a good deal of time discussing and evaluating soils. They see soils as making up a continuum, from dry infertile hill soils to dark, rich valley ones. They see the following terms not as separate types, but as labels for sectors of the continuum. Thus, their discussions of soil often involve questions of whether a given plot of land is *chak lu'um* ("red soil") shading toward *k'aan lu'um* ("yellow soil"), whether some of it is more like *k'aan 'eek'* ("yellow-black soil"), and so forth. Every field presents a continuum from better to worse soils, and this continuum is constantly being monitored, evaluated, and judged. The effect of burning on soil is also known and observed. Typically, on local soils, burning makes the soil more open textured and gritty. It can turn the surface of a clay into a better-drained, more friable substrate. Of course, the effects do not go down very deep, but they can improve the critical layer where seedling maize establishes itself.

This system can be compared with the similar, but far from identical, scheme found in nearby Xocen (Terán and Rasmussen 1994:140). Xocen recognizes a *ch'ich lu'um*, a term unknown in Chunhuhub. Xocen reports a delightful alternative name, *k'amas lu'um*—"termite soil"—for chak lu'um. Possibly "termite soil" really means soil worked over by termites; Chunhuhub does talk of ant and termite debris (used for compost) in similar terms.

Soils in the area are largely chak lu'um, "red soil," that is, rendzina—reddish soil developed from limestone. This soil is the result not only of weathering and dry forest, but also of thousands of years of cultivating and burning. It is, in short, something of an anthropogenic soil. A variant is k'aan lu'um, "yellow soil," a paler and less fertile sort. It is hard to compare these soils with the "natural" soil of the hills and uplands, because there are no "natural" areas to look at. Long-uncultivated areas reveal a deeper, darker, more loamy soil, *chak 'eek' lu'um* ("red-black soil")—but so do the small alluvial valleys that are heavily cropped. In any case, this soil is of superb texture for agriculture, and is high in many useful minerals, but is usually low in nitrogen and phosphorus. To complete the pattern, there is *k'an 'eek' lu'um*, a dry, rather infertile, but darkish soil that develops under high forest on dry but relatively level uplands.

Many of these soils are anthropogenic (Tuxhill et al. 2003), especially the darker, richer ones.

Some steep hill slopes are reduced to *tsek'el*, a pale lithosol—basically not much more than limestone bedrock with small pockets of soil. *Tunchi'* is the extreme case: essentially bare rock. The pockets of soil, at least, are usually dark and fertile, because they have been centers of wild plant growth for long periods. *Sajkab* is limestone, grading from sandy or sticky rotten-rock into solid rock. A variant of this word, *saskab,* is generally used for rotten limestone mined for road surfacing, and a quarry for it is sometimes called a sajkab but is more often known by the mestiza-Maya term *sascabera*.

Chunhuhub lies precisely on the line between hills and flatlands. At its west edge, rounded but steep limestone knobs rise suddenly. Their soil is thin, ranging from bare rock through tsek'el to thin but fertile chak lu'um. They are covered with dry thorn-forest, more typical of northwest Yucatán state than of Quintana Roo. Among their spiny denizens are acacias, cacti, and greenbriars. Perhaps the plant that best sums up the environment is the indescribably thorny *piim* (or *pi'im, Ceiba aesculifolia*)—a tree entirely covered, from root to twigs, with huge savage thorns as thick at base as a man's finger. Among these thorn trees are surprisingly beautiful ferns, philodendrons, epiphytes, orchids, and the little foliage-tufts that temperate-zone residents know as "house plants." These are gathered and exported. Many a leafy resident of a New York or San Francisco window started life on a Quintana Roo hill.

In spite of the forbidding vegetation, these hills are heavily cultivated. Their light, well-drained, friable soils are excellent. Most of the vegetation (by number and size of plant) belongs to the legume family, and typically these fix nitrogen, though some do not. The soils are thus rich in that often-scarce nutrient. Constantly renewed, the soils have no time to lateritize (of which more anon). There is much bare rock on the hills, but what is not bare rock will produce a phenomenal crop of corn or beans.

Moreover, among the hills are countless little and big hollows, similar to the "cockpits" of central Jamaica. These range in size from a few millimeters to half a kilometer across. They trap good soil and are ideal for planting. One can see corn and squash, planted with no attention to neat rows, flourishing with riotous success on what appears to be an expanse of naked white rock. Each plant is growing in a perfectly chosen hollow—small round ones for corn, medium-sized ones for beans, large deep ones for squash.

Some 25 percent of Chunhuhub's 14,330 hectares (ca. 32,000 acres) is hill land. Five percent is fertile valley land at the base of the hills—the mainstay of Chunhuhub's agriculture, when it is not ruined by hurricane flooding (as it was in 1987 and 1995). In these areas—including small pockets within rougher parts of the level lands—are found the chak 'eek' lu'um and *ya'ax jom* "(soil of) green hollows."

The rest of the ejido is flatwoods: forest on a level limestone shelf stretching east to the sea, with soils ranging from chak lu'um to k'an 'eek' lu'um. Particularly common in wetter, flatter areas is *k'ankab*, a poor-quality, clayey, yellow-red soil—the result of too much milpa-cutting on fragile ground. Other soils include a sticky red lateritic clay, also a sign of overuse in the past.

Higher, better-drained parts of this shelf can be as fertile as the hills. Often, however, the soils of the flatwoods are older and have suffered from tropical weathering. The limestone erodes away, leaving whatever impurities (such as iron-rich clays) were in it originally. These impurities lateritize: under the sun and rain, they become enriched in iron and aluminum, oxidized to high and stable levels. These soils are a nightmare of gluey clay when wet, and rock-hard and infertile when dry. They produce a miserable harvest, and are usually (and sensibly) left to produce forest.

The forest of the flat shelflands is anything but primary; it has been affected by cultivation and logging for at least five thousand years. It has found an accommodation of sorts with the human species and with Maya culture. The Maya have been using it so long that they have been a factor in its evolution as a vegetation type. It is meaningless to speak of the "natural" vegetation of Quintana Roo; humans have been affecting the vegetation for about as long as the current climatic regime has existed (i.e., since just after the Pleistocene) and agriculture has been a major part of the environment for three thousand to five thousand years, since post-Pleistocene climates and sea levels more or less settled down (see Anderson 2003; Gómez-

Don Felix (l) and Don Jacinto Cauich (r) in an ak'alche', *a wet area, filled with water during rainy periods, usually damp with pools or wet mud the rest of the time. It grows a distinctive vegetation, often dominated by* 'eek', *the "logwood" formerly exploited for black dye.*

Pompa et al. 2003, passim). The forest throughout is enriched with useful trees and also with nitrogen fixers (see below).

Lowlands are dominated by *'eek' lu'um* (also *box lu'um*), "black soil." This ranges from extremely fertile to extremely infertile. The extremely fertile ones—alluvial lands by wetlands—are apt to be called *'eek' lu'um*, while the infertile—aluminum-enriched, acid soils in sink areas that do not drain well—are more apt to be *box*. (*'Eek* and *box* both mean "black," but the former implies a shiny pure black, while the latter is more properly "dark." Often, *'eek* is a term for "good" black colors and black things, while box is used for things more negative in connotation.) The well-drained, deep, dryish, loamy sectors of thc lowland forest soil universe are, again, chak 'eek' lu'um. Still darker, alluvial-enriched loam is *kaajkab lu'um*. As one descends into marshy or soggy land, this grades into ya'ax jom—places that stay green even in very dry periods. This soil is black, deep, and very rich in organic matter. Sometimes it is little more than the decayed stems of wild cannas and other vegetation typical of ya'ax jom sites. The infertile, acid soils occur in the ancient *ak'alche'*, the wetlands that do not drain and do not receive significant alluvial input. This soil is worthless for cultivation.

In the worst of the aguadas or ak'alche', these soils are clayey, overenriched in metal ions, and almost worthless for farming. They do, however, have one redeeming virtue: about the only thing that will grow in them is a scrubby and murderously thorny tree that happens to produce one of the best natural black dyes in the world. This plant is called *'eek'* (black) by the Maya, *palo de tinte* (dyewood) in Spanish, and "logwood" in English. In times past, it sustained a huge export industry, especially in Campeche. Logwood was cut, especially in the eighteenth and nineteenth centuries, for export to Europe, where dye was extracted (see chapter 8). Chunhuhub was too remote, and too poor in ak'alche' lands, to have been affected, but logwood was cut in many neighboring areas.

Most of the lowland soils are left to forest or cropped on very long cycles. In neighboring areas, some have been converted to cattle pasture, but they do not work well for the purpose. The grass does poorly and the forest keeps pushing its way back in. Moreover, the natural savannahs just to the south produce cattle successfully enough to make ranching a bad option in the Chunhuhub flatwoods.

Overall, Chunhuhub's natural vegetation is about 25 percent dry hill forest, more than 70 percent flatwoods, and 1 or 2 percent ak'alche'. By contrast, Presidente Juarez is a bit more than 50 percent hill forest, almost 50 percent ak'alche'; this was largely cleared for agriculture, but the agriculture failed and the land became rough pasture. Agriculture is virtually confined to the hill forest and to the fertile valleys of the transition zone. These make up 5 percent of Chunhuhub's land. They were, until recently, tall flatwoods with many ak'alche' mixed among them. They are now intensively farmed, except for the ak'alche', which remain too clayey and oft-flooded to be of much use. (One farmer planted hybrid maize in an ak'alche' in 1995. Hurricane Roxane ensued. The day I wrote this paragraph, in 1996, I watched ducks eating the few grains that remained in the lake that was once an ill-planned field.)

Currently, the forest is a mosaic of patches of varying height and history. Some are recent regrowth—milpas turning rapidly back to woodland.

Others are tall forest—what the Spanish call *selva alta* and the Maya call *kaanaj k'aax* (both terms mean exactly the same as the English). This can be a true tropical-postcard selva, with huge trees held together by lianas and covered with epiphytes.

Old-grown forest grows on flat lands with rocky, poor soil. At present, clearing and logging has reduced most of it to medium height, 20 to 30 meters. Old ya' (sapote) and mahogany trees tower to 40 meters, their trunks often 2 meters thick. Almost as large are old specimens of *cheechen, chakte'* and *k'aanchakte'* (*Lonchocarpus* spp.), *ya'ax'eek'* (*Pithecellobium* spp.), *xuul* (*Lonchocarpus xuul*), and *kitanche'* (*Caesalpinia gaumeri*). In early spring the *amapola* (*Pseudobombax elliptica*) raises tufted flower heads. Later, *jo'ochok'* (*Nectandra* spp.) and *pichiche'* (*Psidium sartorianum*) explode in spectacular white displays that scent the air for miles. The *ja'asche'* trails its long tentacles of bloom. From summer into fall, red flowers of the mistletoe glow like neon lamps on many trees (see Anderson ms. for full description of this forest type).

Trees have character of their own. The first one learned by the sensitive outsider is the *chechem* (*Metopium Browni*), an overgrown poison ivy, a poison ivy that can be a meter and a half thick and 30 meters in height. It causes a terrible rash in susceptible people; most of the Maya seem immune. Much more pleasant is the *ja'abin* (*Piscidia piscipula*)—the commonest and most widespread tree of Chunhuhub, if not of the whole peninsula. A large leguminous tree, it has rock-hard wood that lasts forever in house posts, railroad ties, animal pens, or fencing. Posts can be recycled from houses many decades old, and show no damage. Ja'abin flowers spectacularly in May; a good year's display could almost be seen from the moon. Announcing the imminence of the rains by this flowering, it finds a place in Maya rain-bringing ceremonies (Flores and Kantún Balsam 1997).

One next learns the precious woods: mahogany (*Swietenia macrophylla*; *caoba* in Spanish; the Maya name is no longer used) and *cedro* (Maya *k'uj che'*; Spanish cedar, *Cedrela odorata*—neither Spanish nor a cedar, but a soft-wooded mahogany), which are the mainstay of the lumber economy. Also important is the sapote (*ya', Manilkara sapota* or *M. achras*, formerly often known as *Achras sapota*), whose fruit is edible and whose sap is the source of chewing gum. (Today, artificial gum exists—inevitably—but the natural product still finds a good market.) Huge old sapote trees are scattered thickly through the forest; few will cut a sapote, because of its value. Because of this, some 15 percent of the tree cover of local old-growth forest is sapote (figures supplied to the author by personnel of Plan Forestal). The sapotes give an idea of what the forests would look like without constant human pressure. Many are almost two meters thick and 40 or 50 meters in height. In isolated areas, spider monkeys still cluster in them, living on the fruit. Other trees rarely reach anything like such a size; they are cut for milpa or for timber. There are limits on the size at which valuable trees can be taken: 55-centimeter diameters at breast height for mahogany and cedro, 40 for the others. As soon as a tree reaches that size, it is logged. Many are illegally taken long before.

There is little doubt that all of the land in Chunhuhub and neighboring ejidos has been cultivated at some time or another, but some of the flatlands have not been cut or cultivated since the Caste War. The hill forests and the hill-foot zone include some old growth, but are mostly young second growth or active fields and orchards. The fields, of course, are overwhelmingly maize. Beans and squash are intercropped, but are far less important. The orchards are primarily sweet orange, with a substantial acreage in mango, banana, and papaya as well. Cattle pasture does not yet exist in Chunhuhub; the few cattle feed around the edges of fields.

Thus, most of the Chunhuhub area, and indeed most of the Zona Maya, is covered with second-growth forest ranging from five to a hundred or more years old (see, for the general case, Edwards 1986; Tuxhill et al. 2003). Typically, it is between eight and fifty years old. Much of what little primary forest remains has been selectively thinned, which makes it approximate the second-growth forest closely. The thinning creates many light gaps, and here the secondary forest trees thrive; most of the primary-forest ones prefer to start in total or partial shade.

Land that is cropped frequently—that is, the more accessible, fertile areas—becomes dominated by a very characteristic assemblage of trees, dominated by leguminous species. Commonest are *tsalaam* (*Lysiloma bahamensis*) and *ja'abin*, and the unrelated *chakaj* (*Bursera simarouba*). The domination of nitrogen-fixing leguminous trees on recently abandoned land is evident; they colonize soil that has been bared, cropped, and allowed to regrow once its nitrogen is depleted. Such forest will often be broken by large emergent sapote trees. These have been left during previous clearing, because of the value of their fruit and chicle. *Xa'an* (*Sabal* palmettos) and other palms are also left in clearing and become more common over time, especially in moist or hilly areas. Other useful trees are occasionally preserved as well.

The second growth following milpa cultivation takes on a broadly characteristic, simple form, but is infinitely complex in practice. Moreover, it is rapidly changing, as introduced weeds become established. Sow thistle (*Sonchus oleraceus*), for instance, was not observed in 1991, was widespread though rare in 1996, and was reasonably common in 2001. The Maya did not have a stable name for it, which suggests it was indeed a recent invader; people were not really familiar with it. Pasture grasses have spread from cattle ranches and become a terrible problem locally, since they carry fire. Johnson grass (*Sorghum halapense*), already a pernicious weed of permanent cultivation in 1991, had spread considerably by 1996, and still more by 2002, taking over great areas of irrigated land and spreading out into the forests.

The Maya of the area recognize two stages of milpa regrowth, in contrast to the Maya of Yucatán state, who recognize more than two stages.

First comes *sak'ab*, in Spanish *cañada*—milpa in its first few years of regrowth, when it is covered by grass, weeds, and bushes. (Bricker et al. 1998:241 record it as "second-year cornfield"; it is that, but can be a briefly fallowed field too.) Sak'ab is usually dominated by plants of the families Fabaceae (legumes), Asteraceae (composites), and Convolvulaceae (morning-glory vines). Amaranth, pokeweed, and many other familiar weedy groups also occur. In January and February, a sak'ab field is a spectacular sight. *Taj*—sunflowers four to five feet high (*Viguera dentata*)—make it a blaze of pure gold. Bees swarm on these *tajonales* (taj fields), producing the honey that is one of Yucatán's major exports. Morning glories of a dozen hues vie with the huge sweetpea blooms of *jeeret* (*Centrosema* spp.) and other leguminous vines. Above and behind these, regrowing *kaskat'* trees are covered with white flowers as big as saucers, and the star-yellow flowers of the *ch'ooy* are the more spectacular because of that tree's leafless winter condition. The dry season, coming in March, quickly turns the sak'ab field dull and brown. A sak'ab grows into *huamil* (a Spanish term, originally a Nahuatl word): second growth with small trees. Huamil consists mainly of regrowing trees, especially those that have been deliberately left (especially xa'an) and those that enthusiastically resprout from cut stumps (especially ja'abin, chakaj, kaskat', and kitanche'). Huamil has its share of vines, but they are usually woody ones like *saya ak'* (*Vitis* sp.), not the soft annuals of the sak'ab. A few trees seed themselves thickly into huamil; acacias and mimosas of many species are particularly prone to invade as seedlings.

Huamil (especially near town) is subject to constant thinning for poles. These are most often for use in house walls and fences. Roofs, digging sticks, firewood, tool handles, and various other uses all make their claims. This leads to a selective removal of trees that are hard and straight when young. It also is good for the forest, providing a useful thinning. Firewood collecting eliminates wildfire danger by removing burnable windthrows and dead timber. Firewood collecting is also highly selective: ja'abin is preferred, with *ts'its'ilche* (*Gymnopodium floribundum*) coming second. These woods burn very hot and clean, and are dense enough to give considerable heat for a long time. Charcoal is not made in Chunhuhub today; it once consumed much wood, especially chakaj, which grows fast and produces a lot of good charrable wood in a short time. In some parts of Yucatán, one can still see chakaj trees coppiced high in milpa cutting, so that they will regrow quickly and replenish the charcoal supply. Fruit collecting and other uses also affect the regrowth of the forest. Intensive collection of medicinal plants is constant, though normally too diffuse to have much effect. Moreover, huamil is highly floriferous and produces much of the honey of the area. A huamil is a highly managed environment, producing a vast wealth of resources and growing into a forest whose composition is shaped by that resource production.

Maintaining Maya lands in permanent forest is neither necessary nor desirable. Without Maya firebreaks and fields to interrupt wildfires, the hills would almost certainly burn frequently and regularly, in huge holocausts that would destroy all trees and would lead to rapid degradation of the land. When wildfires swept the less populous parts of Quintana Roo after Hurricane Gilbert, the fires stopped when they came to the "managed mosaic" (Fedick 1996a).

CHAPTER THREE

Agriculture

CHUNHUHUB, and the whole of central Quintana Roo, is based on agriculture. Subsistence cultivation of maize and its traditional associates is by far the most important activity.

The agriculture of the Zona Maya of Quintana Roo is still in the hands of independent Maya cultivators, owning their own land and communities, and holding and working land in common. The ejidos make available shares of their common lands to families (nuclear or extended); the family is the unit of work. Wage labor is usually no more than a way of supplementing a lifestyle based on farming and forest use.

Michael Kearney (1996) has pointed out the misleading nature of the word *peasant* as applied to Mexican farmers. *Peasant* was and is technically a term for farmers under the Old Regime in Europe—people in a special, and low, class, with few rights and many obligations. Such a class, as a formal entity, has not existed in Mexico since the nineteenth century. Mexico's ordinary rural people, including those of the Zona Maya, call themselves *campesinos*: "countryfolk." It is a label that includes farmers and farm workers, small traders and village men-of-all-work, craftswomen, and skilled laborers. Translating it as "peasants" is mistaken and patronizing.

Today, Mexico's campesinos are usually either independent yeoman farmers or rural proletariat. The Quintana Roo Maya, heirs of an ancient civilization, have preserved their independence and many of their traditions while adapting successfully to the modern world. Neither fully "traditional" nor reduced to the impoverished state of many Mexican rural people, they are developing a range of accommodations to the rapidly changing conditions of southeast Mexico.

Farming families receive from the ejido one to four hectares (2.4 to 9.6 acres) of milpa—slash-and-burn fields under shifting cultivation—in any given year, and also have large dooryard gardens. Serious farmers with families to support take the full four-hectare plot to which they are entitled by ejido rules; others with less demand will take one, two, or three. The milpas are moved every two or three years; the dooryard gardens are stable, and often contain old trees.

Since milpa work is exceedingly hard and produces little of monetary value, there is every incentive to use no more land than one needs to feed the family. Maize is low in price and yields only about a ton per hectare (on very good soil, and with exceptional luck, a ton and a half). Milpas produce beans, squash, and some other crops. Most of these are raised in the *pachpak'al*, a small central area where various crops are raised; tomatoes, chiles, and young orchard trees are favorites, but some people raise up to thirty or more species.

Today, modern orchards, vegetable farms, cattle ranches, and small livestock operations have been added to the traditional pattern. Chunhuhub is typical of Quintana Roo ejidos in having some 5 percent of its cultivated land irrigated.

Chunhuhub, like almost all Maya towns, is deeply devoted to slash-and-burn agriculture. This is not just a source of the staple food; it is a religious act and a moral duty. A man who does not cut milpa was, until recently, not fully a Maya (see Redfield and Villa Rojas 1934; Terán and Rasmussen 1994). Today, life in a modern town has made storekeeping, teaching, road work, and other occupations acceptable and prestigious. People in those occupations have little time for *meyaj kool* (working milpa). However, almost everyone who has a reasonable amount of available time still cuts at least a small cornfield and garden.

Mexico's 1994 agricultural census recorded 6,611 farms in the area, and 5,123 of them were solely for subsistence (INEGI 1994:73). Only two farms reported export sales. The vast majority of the farms (5,381 of them) were under 5 ha (INEGI 1994:154). Maize is by far the dominant crop, accounting for three-quarters of crop acreage but only a quarter of commercial crop value. Citrus accounts for most of the rest; watermelons and vegetables are also important (see INEGI 1995:176).

The ejido system harks back to pre-Columbian collective rights and to Colonial village tenure, but its actual form is a product of the Mexican revolution. Ejidos were granted from expropriated estates, or, as in central Quintana Roo, from land held under traditional title but legally owned by the state. The ejido grants of the Zona Maya were made in the 1940s and 1950s, or, in some cases, later.

Under the system that existed up until 1993, ejido land could not be divided, sold, or alienated. It was still national land in some ultimate legal sense, though the ejidatario families could do what they wanted with it. In 1993, Article 27 of the Mexican Constitution eliminated this system. Ejido grants were released to their inhabitants, to maintain, divide, or sell, as they saw fit.

Chunhuhub, like almost all the Maya ejidos, has resisted the temptation to break up into privatized parcels. Since the alteration of Article 27 of the Mexican Constitution, about a third of Mexico's ejido land has been parceled out, but not even 1 percent of Quintana Roo's has been (INEGI 2000d:1:376–77; Yucatán and Campeche have parceled out about a fourth of their ejido lands; at the other extreme, Veracruz's are almost all privatized). New landowners are all too often plied with alcohol and persuaded to sell off their lands for a pittance. (The same fate often happened to Native Americans in the United States when reservations were "allotted.")

In the Zona Maya of Quintana Roo, however, the Maya have been communal and careful, and preserve their lands as collective holdings. Land is allotted to families, according to what they want and need. If they are making milpa, they are allotted land on a one-year or two-year basis. If they are planting an orchard, they get a longer-term allocation: twenty-five years or more on paper, indefinitely long in practice. Until recently, the local ejidos charged no rent. Recent immigrants (who are not ejidatarios) get land along with everyone else, since land is not yet in short supply. Today, new immigrants have to pay for long leasehold, and the price is close enough to local land sale prices that leasers simply use the term "buy" for this transaction.

However, those who are not ejidatarios can do this only at sufferance of ejidatarios. Moreover, non-ejidatarios cannot get land when an ejido is divided. And ejidatarios get preference for the choicest spots—irrigated land and the like. Finally, daughters do not inherit ejido rights unless they are full heads of households when their fathers die. Thus widows often find themselves disenfranchised, to their hardship.

Chunhuhub's ejidatarios do not all plant in a given year, even when it is land they regularly work. Non-ejidatarios often lease such land and plant it.

Much of the orchard land is now de facto privatized; it is still theoretically leased, but is openly bought and sold. A hectare of young orchard, for instance, sells for 3,000 pesos (1996). By 2001, it had become possible for non-ejidatarios to buy—de facto—forest for milpa cutting for 500 pesos per hectare. Technically, they are acquiring only lease rights, so there might be future problems of tenure.

Maya preservation of communal landholding is partly a political move. The Maya of Quintana Roo are fiercely protective of their independence and of the land that is its obvious basis. They still have strongly communal values. They are also highly suspicious of government attempts to split up ejidos and permit sale of land to outsiders. Many see the government's attempt to privatize ejido land as a ploy to destroy Native peoples' solidarity, shatter their economy, and reap the benefits by taking their property. This viewpoint is perhaps especially common in the Cruzob towns such as Tusik and Señor, but in Chunhuhub, too, it has its defenders.

However, the main reason for preserving the ejido system is economic. Slash-and-burn cultivation is still the major activity, and it does not lend itself to parceling out land in small holdings. Every family needs to relocate its milpa every two or three years. In these tight communities, every farmer wants to have a chance at the very best plots, and year-by-year allocation allows this to happen; people take turns getting the best or being stuck with the worst. This means that a huge area of land must be available. Moreover, in a wet year, everyone will want dry, well-drained land, and in a dry year they will want moister sites. Under this economic regime, there is every reason to hold land on a community basis. Also, forest management can be done only on an ejido basis, for reasons that will become obvious.

For all these reasons, by 2000, only one ejido in the area had divided its holdings, and even it has not sold them off or alienated them. It was

Emiliano Zapata, a small, uninhabited ejido (its ejidatarios live in Chun-huhub). Its twenty-five families could not agree on any issues of collective management. They thus divided the ejido into twenty-five blocks. This was done with open and expressed regret, as a last resort, and has indeed led to very bad management (according to local reports relayed to me). The other local ejidos point to it as a horrible example of what can happen.

However, privatization is gaining. Presidente Juarez allocated its ejido lands to families in a gradual process, and as of 2002 was largely privatized de facto (though not de jure). This was done because some families mis-used the forest, burning too much too often.

In addition to ejidos, there are two other classes of land in the area: government land (mostly federal) and private land. The former is primarily fairly remote or worthless land, largely thorn-forest on rocky ground. It is not much used. The latter is divided into *ranchos*. These vary in size and seriousness of cultivation. At one extreme are casual, relaxed little villas, often no more than weekend places for ejidatarios or shopkeepers whose economic interests are elsewhere. At the other end is the huge García ranch, Rancho el Corozo, operated according to the best principles of land management that can be brought to the remote Quintana Roo bush. The Garcías run a diverse operation, with modern commercial agribusi-ness, sophisticated cattle ranching, orchards and timber plantations, and remote forest managed for sustainable yield—to say nothing of plans for ecotourism on the plantation's lake (a large natural body of water of travel-poster loveliness). Several other large ranchos have experimented with new technologies and ideas, as well.

Milpa is a Nahuatl word for the field system known in Maya as *kool*, in Spanish as *roza-tumba-quema* ("slash-fell-burn"; RTQ for short), and in English as "slash-and-burn" or "swidden" (see Conklin 1957; Spencer 1966). Maya milpas have been particularly well described in the literature (esp. Terán and Rasmussen 1994; Terán, Rasmussen, and Cauich 1998; Várguez Pasos 1981, a collection of classic articles). Agriculture is sus-tainable, management is good, and the Maya know and understand this (Anderson 2003, ms.; cf. Atran et al. 1999; Fedick 1996, Remmers and de Koeijer 1992; these sources inform all that follows in this chapter). Much of this is maintained by social control systems, often religiously constructed (see chapter 6). The dooryard gardens are also an important and thoroughly traditional component of the system (Anderson 1996b; Anderson 2003; Herrera Castro 1994).

Farmers carefully plan their slashing and burning cycles. Slashing may begin many months before sowing is planned, since cutting the larger trees is a slow process. (Only the very richest can afford chainsaws.) Many of the rain-forest trees have wood hard enough to turn an axe blade, and some of these grow two meters thick. They often must be ringbarked and left to die, then burned in place. This has the benefit of creating many hollow trees, which then become the homes of wild bees and other useful wildlife. Hollow trees are a well-recognized and, sometimes, a deliberately planned side benefit of cutting milpa.

Maize remains *the* Maya crop. It traditionally provided 75 percent of Maya calories (Benedict and Steggerda 1936; Peraza López 1986; Steggerda 1943; White 1999), and still provides 75 percent for the most traditional cultivators in the Chunhuhub area. In 1997–98, 80,181 ha maize were recorded in Quintana Roo, of which 20,289 were recorded in Felipe Carrillo Puerto *municipio* (INEGI 2000c:73). In fact, this is a huge undercount, since nobody bothers to count all the subsistence milpas. Essentially all of the maize agriculture is rainfed. Oranges were planted on 3,866 ha at that time in Quintana Roo. Some 1,215 ha were in Felipe Carrillo Puerto municipio; most of this acreage was irrigated. The municipio reported 13,592 tons of maize harvested, and 2,256 tons of oranges. Felipe Carrillo Puerto reported no other consequential agricultural activity (INEGI 2000c:73, 75).

Maize varieties include various local ones, loosely divided into quick-maturing corn (mejen, "young"), which ripens in two and a half to four months, and ordinary local corn (*xnuk nal*, "large corn," or *nalt'eel*, "rooster corn"), which takes four to six. These categories are vague and general, and apt to blend into each other. (Nalt'eel, when used for all local corn, is a broader category than the formal botanical category that crop scientists have named Nal-tel, from the Maya word. However, some Chunhuhub farmers use the term to apply to something like the botanists' Nal-Tel, and use *xnuk nal* to refer to local large corns in general.) The local corn includes a fantastic variety of types—white or yellow, dent or smooth, hard or soft, quick ripening or very slow, tall or short. The classic nalt'eel types hold their own, but many others, named *criollo* (creole, local), are grown; these seem to be migrants from elsewhere, or hybrids of nalt'eel with foreign strains. Names are vague and shifting. An excellent recent study by Luis Arias and his associates (2003) has recorded almost two hundred names for maize and maize products in the nearby and culturally close community of Señor. Arias's group observed and collected a truly awe-inspiring variety of maize ears—a spectacular spread of colors, shapes, and sizes. Most popular there was a slender white-kerneled ear known as *bakel* ("bonelike"). It makes delicate, rather soft tortillas. This variety is known, under the same name, in Chunhuhub (it is considered to be one of the criollo races). It is less popular than thicker, yellow-kerneled maizes that make substantial, chewy tortillas.

Modern commercial hybrid varieties are widely used, but are not good choices except on the irrigated, mechanized land. The native maizes are hardier, and thus yield as much as the hybrids do, unless conditions are exceptionally good (fertile soil, some fertilizer, good luck with the rains, and/or similar factors). The standard nalt'eel and mejen are tough, deep yellow, and not very sweet, like hybrid corn. Tastes have adjusted accordingly, and there is no price differential in favor of the local corn, as there is (for instance) in much of Oaxaca.

Beans (*Phaseolus vulgaris*) are not very important in the economy of Chunhuhub. Traditional families and less affluent individuals eat them almost daily, but often in small quantities. Many families prefer buying them to raising them. Beans, especially the preferred black beans, simply do not grow well, and it makes more sense to invest one's time in other vegetables or in saleable crops. Beans cannot be produced in sufficient quantities to be a serious economic enterprise. Sieva beans (small Mexican lima beans, *P. lunaris*), black-eyed peas (*Vigna unguiculata*), and pigeon peas (*Cajanus cajan*, locally known as *lentejas*) all do better than black beans, but people rarely bother to grow them, though they could add much to food security and quality. Beans tend to be considered poverty food; extra capital and effort therefore usually go to other things.

Tomatoes (*Lycopersicon esculentum*) abound, both for home use and for sale. Chiles are universal and vitally important in the domestic economy. There are many varieties of common chile (*Capsicum annuum*), including sweet bell peppers as well as hot varieties ranging from the mild *xkatik*

Sowing: a hole is made with a stick, and three or four maize kernels are dropped in.

to the burning *jalapeño*. More and more varieties are entering the area, as people grow them commercially for the cities. Grown for home use is the *maax* or Tabasco chile (*C. frutescens*), which is liked for its food quality but even more for its near indestructibility; it becomes a large woody bush that will live under the worst conditions. This is useful when one's garden is a limestone rock. Particularly popular for food is the habanero (*C. chinense*), whose name betrays its post-Columbian arrival from Cuba. It is originally a South American chile. Its incandescent hotness made it successful, and one cannot imagine Yucatec Maya life without it. Old Maya farmers munch habaneros one after another, like apples. Larger-scale farmers like the Garcías sometimes grow both tomatoes and chiles in substantial quantities, for commercial sale. Whiteflies carry damaging viruses, but less so than elsewhere in the Yucatán; Chunhuhub farmers do not spray heavily, and thus do not turn the whiteflies into a plague by killing their predators. The Herdez company (a major Mexican chile packer) leased land and successfully grew chiles one year, but, for reasons not known to my friends, did not return.

Various squash are common (as is true everywhere in Mexico). Squash are of several species: *Cucurbita pepo, C. moschata, C. maxima* (?), *C. argy-*

rosperma (a.k.a. *C. mixta*), *Sechium edule* (*chayote*). Squash are grown not only for flesh but also for the seeds, which are ground whole to make the popular seed meal called *sikil*. Varieties containing little except large seeds grow specifically for this, but small seeds from fleshy varieties will serve. Bottle gourd, *Lagenaria siceraria*, once grown very commonly for bottles and dishes, is being replaced by plastic.

A few families raise radishes, cilantro, kohlrabi, cabbage, and other vegetables for sale. Most who have gardens raise the first two for household use; both crops are extremely popular for relishes. Every family with a food-producing garden grows garlic chives (*Allium tuberosum*), using *k'aanche'* (raised platforms on legs), pots, or buckets to keep them out of range of animals. How this Chinese crop got to the peninsula is something of a mystery; perhaps Chinese laborers introduced it in the nineteenth century. Ordinary chives (*Allium schoenoprasum*) do not occur, though they are the typical chive crop elsewhere in the peninsula.

Some farmers grow vegetables seriously and sell them on a local scale. One family practices highly intensive mixed farming and sells tomatoes and other vegetables as far as Felipe Carrillo Puerto.

Fruit is the main income source in Chunhuhub. Watermelons are the most profitable crop in terms of money per hectare or per input, but oranges are the great standby. Almost everyone has a few citrus trees, and most serious farmers have substantial orchards (known as *parcelas*). All these grow both in dooryard gardens and in orchards. Many fruits occur primarily in dooryard gardens, though mameys and avocados appear in orchards too. Mameys (*Pouteria mammosa*), sapotes (*Manilkara sapota*), avocados (*Persea americana*), various *Annona* species including *oop* (*A. purpurea*) and cherimoya (*A. cherimola*), caimitos (*Chrysophyllum caimito*), cashew fruit (*Anacardium occidentale*), coconuts (*Cocos nucifera*), papayas (*Carica papaya*), nances (*Byrsonima crassifolia*), mamoncillos (*Melicoccus bijugatus*), guayas (*uayum, Talisia olivaeformis*), and guavas (*Psidium guajava*) are raised in significant quantities and sold (for the full story see Anderson 2003). Ciricotes (*Cordia dodecandra*), cacao (*Theobroma cacao*), coffee (*Coffea arabica*), and a vast host of additional fruits are raised as experiments or for home use, and now and then sold if they produce any significant quantity of fruit. Some families have sugar cane, *mak'olan* (*Piper auritum*—grown for its large edible leaves) or other unusual perennial crops. Pitahayas (*Hylocereus undatus*), the lumpy, shocking-pink and lime-green fruit of the night-blooming cactus, are a dramatic-looking local fruit. They are native to the area, and have only recently begun to spread world-wide. I recently saw them sold in China, where their appearance has led to their being given the poetically descriptive name of "fire dragon fruit." The Maya name, *wob*, may be just as graphic. It is homonymous with—and perhaps derived from—the Maya name of the giant marine toad (*wob muuch, Bufo marinus*), whose croak sounds like *wob* and whose large, warty body looks rather like a pitahaya.

Front gardens and often back gardens as well host a vast array of ornamentals, native and introduced. All sorts of wild plants grow; most are useful, at least for minor medicinal purposes. Fruits, especially oranges and watermelons, are the mainstay of the Chunhuhub cash economy. These are trucked to urban areas, especially Cancun.

Among rare and specialized crops are medicinal herbs. Most people just gather wild plants, with the exception of the universal rue (*Ruta chalepensis*), mints (*Mentha* spp.), aloe vera (*Aloe* spp.), basil (*Ocimum basilicum*), and the like. However, a few herbal doctors grow a range of supplies.

Marijuana (*Cannabis sativa*) grows in remote, isolated areas. I made it a point to avoid all association with this industry, so have little to say about it, except to note that it exists but is uncommon and of minor importance. In one remote community, a seller could not believe that any gringo would come to his isolated area except to buy the weed, and he was difficult to shake off.

The process of learning to farm is a part of the process of learning about the world. People remain open to new varieties and methods. Don Felix Medina Tzuc, on a rare visit to relatives back in his home town (which he left for Chunhuhub about forty years ago), collected cuttings of local varieties of fruit trees and flowers, and seeds of the hometown varieties of squash and beans; he also observed with interest all the plants and planting methods used in the area.

The beginning of the planting year is announced by the Bright-rumped Attila, a bird whose wild, loud, exquisite song carries a mile over the silent forest. The bird is sometimes called *pak' sak'ab* "plant your fallow field."

By late February, the attila has begun his song, and the "first season" (late winter) has brought its busy days of slashing and clearing. Midwinter is a time of relative ease. Men eat heartily, storing up energy for the steady, hard labor of February and March.

By mid-March, the land's solid green cover is broken by more and more brown squares. A few plots of high forest are down. Men recut many huamil fields and weed the citrus orchards, head high in brushwood and weeds. Roadsides have been burned; the fields will be next. They will dry a short time, before the blazing heat of May makes burning really dangerous.

In Yucatán state and dry parts of Quintana Roo, farmers have to clear firebreaks around fields before the main burning (see Terán and Rasmussen 1994; Várguez Pasos 1981). In most of Quintana Roo, the forest is so wet that it will not burn except in the very height of the dry hot season, so firebreaks are usually unnecessary. However, in dry springs (as in 1991) or in springs following major hurricanes that leave much dead wood in their wake (as in 1990 and 1996), fires get out of control and do serious damage. Agriculture and forestry workers fanned out over the state in 1996, advising farmers on how to take precautions. Chunhuhub has had its disasters. In 1991, a rather dry spring and careless burning led to massive wildfires that damaged considerable areas of forest. In 1996, the workshops were largely successful, but a few intransigent souls burned carelessly, starting small local conflagrations.

After burning, the soil cools under its new coat of ash. People come to clean up branches and make small fires of them. Then comes the day—inevitably one of the hottest, sunniest days of the year—when the family is out in the milpa with digging sticks. The ancient technique is still the best. A sharpened stick is inserted in the ground and pulled to create a wide hole, 5 to 10 cm deep (depending on how light and how moist the soil is). Planters move fast, but not too fast to select the placing of the hole. They seek a spot where the soil is deep and—if possible—protected by rocks, dead wood, or the like. Maya farmers have an ability, born of long practice, to assess the small hollows (*jaltun* and other terms in Maya; *huecos* or *rejolladas* in Spanish) in the limestone hills. Many fields appear to grow from solid naked rock. Close inspection reveals that each plant is growing in a soil-filled hollow. Corn needs much less space than squash; maize roots densely fill a ball of soil of less than a cubic meter, while squash roots go down 2 meters or more and spread out for many meters through the soil layers. Chiles can live in very little soil over rock, while beans need deep, good-quality soil (which is one reason they are little grown, such soil being rare). Thus, pockets are examined, and tested by digging stick and machete, to see how big they are and what size of root system they can accommodate.

Because of these soil difficulties and related problems, the classic Middle American system of intercropping maize with beans and squash is rather rare. (It is common in some other parts of Mayab.) Beans under such conditions usually produce only about 100 to 200 kg per ha. In the Chunhuhub area, intercropping is done only in the richest parts of the milpa, where a real pachpak'al is often created. In most Chunhuhub soils, maize does better than anything else, and it is what people want most. On the other hand, corn-bean-squash intercropping is common wherever the soil is good, and in really rich patches one also finds tomatoes, chiles, lima beans, papayas, and many other crops.[1] Though everyone wants to grow maize, the more intensive phases of milpa cultivation, such as intercropping, are becoming less fashionable. Most prefer to put their efforts into orchard trees. This trend has been notable over the years I have been studying Chunhuhub. The older farmers, who intercropped religiously (often in a literal sense), are retiring, and the young usually prefer to buy their beans. More traditional towns elsewhere in the peninsula continue to intercrop—and to practice the old religious ceremonies—more enthusiastically (see, e.g., the excellent study in Yaxcabá by Tuxhill et al. 2003).

A farmer drops about four kernels of corn into a planting hole and pushes soil over them. Seeding rates run around 10 to 12 kg per ha of maize seed. Beans and squash may go in the same hole or in a separate one. Seeding rates vary from about 10 kg maize per hectare to twice that. The lower figure would usually be in an intercrop, with (as in Don Felix's milpa in 1996) some beans—about half a kg per ha—and six to twelve sweet potato vines.

A typical milpa planting was that of the Gongora family in 1991. They sowed 15 kg of seed per ha. Father and one son planted it in six days, using five to six seeds per hole instead of four. They had saved the seed corn in husk with *cal*—powdered lime made from limestone—to preserve it from the insects. Maize will store two years this way. The biggest, best developed ears went for seed. They sowed some beans, chiles, and other vegetables in the milpa. They also had squashes, sweet potatoes, black-eyed peas, lima beans, and other crops.

The Gongoras offered *santo 'uk'ij* (or, in Maya pronunciation, *saantoj 'uk'ij*, a special maize gruel, like that for church—green corn as well as ripe, with sugar) in the milpa to pray for rain, safety from snakebite, and other troubles. This prayer is to the *yuum* (Maya for "lords" or "gods"—called in Spanish *vientos espíritus*, "spirit winds"). The Gongoras sought hawks as guardians, because they drive grain-eating birds away and eat snakes and pests; hence they offered pozole to the hawks, to lure them in.

The first heavy rains come in mid-May, and by mid-June the regular summer pattern of daily thunderstorms is usually established. The maize grows rapidly. Rain is critical. The dangerous season is May and June. March and April are always dry. People plant a little corn and hope some of it grows. Serious planting is done in late April and early May, to catch the rains that normally begin about May 15. An early-May rain has everyone out, breaking all other commitments, to get their milpas planted while the soil is wet. Often, however, rains are scanty till mid-June or even later. A thin crop results. Maize pollen needs moist air to survive. Drought rarely kills the plants themselves, but it kills the delicate pollen before it can fertilize the tassels. Even a quite brief and mild drought—if it comes in full pollen-shedding season—can devastate a milpa. Corn ears become *t'iich* (not filled out) or have *belwiij*, "hunger roads" (unfilled kernel rows). People resort to omens: snakes hissing or black eagle-hawks calling give promise of rain.

Chunhuhub gets enough rain to make irrigation unnecessary if only the *chaak* gods were more regular and predictable. As it is, the rain usually is too scanty in spring, too heavy in early fall, or otherwise not quite on the mark. In 1995, a spring drought gave way to Roxane's torrential September downpours. In 1996, torrential late rains destroyed crops in low ground. In 2001, a very local but devastating June-July drought totally destroyed the maize crop in much of Presidente Juarez, while a heavy August rainfall destroyed some of what was left. The drought hit the high ground, the rain the low places. Thus both were hurt. On top of that, the weather created ideal conditions for a mysterious disease that yellowed the plants and rotted the growing ears. The result was real want; Aurora Dzib and her family, among many others, faced serious malnutrition. As of old (Feldman 2000), people lived on wild foods and root crops.

The result of all this is that good drainage and irrigation are both desirable—and unfortunately they do not mix. Good drainage is found on the hills, with shallow fast-draining soils. Irrigation is practicable only on the deep-soil flatlands, which often flood in heavy rains.

In 1995, Hurricane Roxane passed directly through the center of Quintana Roo. The actual wind damage was serious enough, but the real problem was the soaking, constant, heavy rains that inevitably follow a hurricane. They flooded fields, soaked the corn ears, and caused massive rotting. Maize that survived was that which was planted on high ground— bitter irony for those like Don Felix who had planted in particularly rich, fertile, moist lowland: they lost almost everything. Hybrid corn, in particular, was a loss. By contrast, many of those who had cut fields on the dry, poor hills west of Chunhuhub were little affected. Arsenio Medina, for instance, planted 4 ha and got 2 tons of maize. He lost the best of it—that planted in the valleys—but still got more than half his expected yield. Domingo Yam was even luckier. He had planted the tough traditional maize in the hills and had finished the doblado before Roxane hit. He lost almost nothing.

Insect control is rarely needed. One can use a smoky candle to smoke ears (doubled or harvested), killing the *bichos* (Spanish for "insect pests"; Maya, significantly, lacks words for "pests" as such). Heavy rain at night in flowering season is the worst danger to maize (after the hurricanes); it is called *choko' ja'* ("hot water," here in the sense of "fierce rain"). Besides directly damaging the plants, it breeds diseases. Sharp drought is a close second. In general, maize is a delicate crop and not ideally suited to the Yucatán climate. Unfortunately, no better grain for the habitat exists, and the really well suited crops—the root and tree crops—have their own problems: lack of nutritional value in the former case, and excessively long maturation time in the latter (a mamey tree requires fifteen years or more to come into good bearing).

The prudent planter will try to have some maize planted in high ground, some in low. Even this measure was futile in Presidente Juarez in 2001. However, yet another protective measure did work: planting some maize late, when rains have *certainly* started. Most farmers plant by mid-June, but in 2001 some Presidente Juarez farmers waited until rains had clearly come on full and strong, and they were saved.

Fruit grows best on lands that are flat and very fertile but still have drainage. Here it can survive floods yet receive the irrigation it desperately needs in late winter and spring. For maximum bearing, orchards should get at least two deep irrigations at this period, and more on dry sites. In fact, pumps are always breaking, and fortunate indeed is the orchard that gets irrigated regularly. Fortunately, mangoes and citrus are drought-hardy, surviving without irrigation if planted in reasonably moist ground. Row crops like watermelons need at least two irrigations in the dry season but survive with one or none if planted in summer for winter harvest.

Weeds flourish. Farmers carry out at least one and usually two weedings. Some, compulsive about cleanness and rich enough to hire cheap labor, do three. Rarely indeed will a family do its own weeding three times over; self-exploitation has its limits. Usually, the process consists solely of cutting down the weeds (including resprouts of trees) with machetes or a smaller, curved knife like a small sickle. Weeds, weeding, and the small curved weeding knife are all called *loob*. Chopping back weeds—or any vegetation—is *chak*. In Yucatán Spanish, weeding is *chapeo* (from the Cuban-Spanish verb *chapear*). A more general process of weeding, pruning, and trail clearing is *limpio* ("cleaning").

Orchard weeding is more or less continuous, one section at a time— about a mecate a day; by the time one has done a leisurely job of chopping down the weeds in one's parcela of a hectare or two, it is time to start over.

Only rarely, when special circumstances create an exceptional need for reducing the competition, do workers go to the trouble of uprooting weeds. Usually, corn grows fast, shading out the weeds. If squash grows, it finishes the job, providing a complete kill—not only do its aggressive leaves shade out everything in their path, but its bitter chemical arsenal of cucurbitacins and other toxins can destroy both insects and weeds with devastating thoroughness—all without harming the maize. Before the shade grows dense, weeding is a frantic activity—a savage war against plants that are tough and usually equipped with ripping thorns or stinging hairs. One has to watch carefully; an eyeful of the hairs of the "black jaguar" plant (*'eek' baalam, Croton flavens,* an infuriatingly abundant weed) can blind.

Many weeds are extremely thorny or prickly, and some of them host stinging ants and wasps. Widely, in ancient Mesoamerica, a myth taught that the gods had made humans out of maize dough and had had to shed their blood on the images to give them life. Maya rulers had to repeat the act, sacrificing their blood on public occasions. Watching Maya cultivation brings this story to life: people today must sacrifice their blood to the corn. No milpa in the Yucatán matures without claiming its share of human blood. It is easy to see the origin of the ancient myth.

By the middle of June, roasting ears are available, and the harvest of early ripening maize comes in July. At some point around that time, people go to the fields for the *doblado*: bending the ears down so that rain will not get in and pests will have a harder time breaking through the shucks. (This pan-Mesoamerican custom has been challenged as unnecessary, but it has real benefits; see Wilken 1987.) The main harvest begins in September and continues indefinitely; people store maize by leaving it on the stalk, and pick it as necessary. Maize ears store very well on the plant in Chunhuhub. Squashes and tomatoes keep ripening over months. Beans store in the pod. People are still actively harvesting in December, and combing the milpas for stray ears and pods as late as February and March. Meanwhile, the main orange harvest comes in—it runs from September into December, with a peak in November.

Meanwhile, clearing for next year can begin. The *le'chak* (initial clearing, mostly just girdling big trees) comes in August. In fall is the more thorough *je'chak*. Finally, in winter (January–March), the full *pui chak* takes place. The whole milpa is cut. Farmers prepare to burn it as soon as it dries.

Root crops are the back-up staple if all else fails. Milpas in low, wet ground may lose all maize to flooding, but jicama (a tuber-growing bean, *Pachyrrhizus erosus*; the English/Spanish name comes from the Maya *chiikam*) and manioc survive. The Maya point out that, in the bad old days, such root crops, especially sweet potatoes, were the only food after a hurricane. In fact, they were for many people even in 1996; Don Felix and his family lived mainly on sweet potatoes for months. Elders remember much more serious problems after Hurricane Janet passed in 1955. At that time, western Quintana Roo was still genuinely wild. People had only their subsistence crops. Janet wiped out the maize. People lived on chaya and roots till the next harvest came in. Don Pastor remembers his grandparents making tortillas of manioc and jicama, and harvesting the wild jicama (*kuup*) as well as wild greens such as the bitter shoots of the *jujuub* tree. Today, the young, who are forgetting much of the old lore, are still made to learn famine crops. One never knows. Even in the modern world, true catastrophes, like the one that struck Presidente Juarez in 2001, do happen.

The other link in the traditional cycle is the dooryard garden. Originally, this seems to have shared the name *pachpak'al*, but now it is called by Spanish names, *jardin* or *solar*. I have described dooryard gardens elsewhere (Anderson 1996b, 2003) and will not do so here (see also Caballero 1997; Gómez-Pompa 1987; Gómez-Pompa and Kaus 1990; Restall 1997). Dooryard gardens have been important to the Maya since at least the Classic Period (Killion 1992; Sheets 1998; Sheets and Woodward 1997). Some Classic sites may have been real garden cities (Chase and Chase 1999).

Fertilizer is still a rather rare commodity in Chunhuhub. Few crops are profitable enough to be worth it. Commercial vegetables and watermelons are most apt to be fertilized. Young trees and the like sometimes get sawdust mixed with dung. The dooryard gardens fertilize effortlessly; they are everyone's outhouse and garbage dump, and they are full of animals whose droppings add to the rich soil. Studies in Yucatán and Quintana Roo consistently show high nitrogen and mineral levels in these soils (Ottilia Baer, personal communication 1989, 1991).

Worst pests are the large mammals: peccaries, pacas, deer, coatis, raccoons. The peccaries, pacas, and deer are now almost gone, because of overhunting, but within recent memory they could be serious. (Pacas were particularly fond of squash; the other two preferred maize.) A herd of peccaries could devastate a milpa. Coatis and raccoons are still a problem. I saw a family bringing home a dead coati one day. I commented: "I see you have a coati there." They answered: "It was eating our corn, so we're going to eat it." On another day, Don Felix and I met one of his nephews, carrying a shotgun, staking out his milpa—a coati had been raiding it and had eaten a good deal of his maize. A few hours later, he came in, triumphantly bearing the coati and talking of what a good *makum* (stew) it would make. As these stories suggest, hunting is the control measure for mammalian pests. Meat is a major goal. If the animal is not worth eating, it is not usually enough of a problem to be worth hunting. There is usually enough in the milpa for the animals to have their share—and traditional Maya farmers are astonishingly generous about this. Countless times I was told that the animals are living things, too, and need their food; God created them, and they are delightful to have around; we can spare a few ears of corn. Parrots and grackles take a toll of cornfields, but they are not abundant and they are highly mobile, so their toll of any one field is rarely significant. One pest control device is the *k'aanche'*, a raised platform for seedlings.

Insects and diseases rarely constitute a serious threat. A wet year creates problems with corn smut, but conditions are never really optimal for this fungus; hot sun inhibits its growth. (Unlike central Mexicans, Yucatecans do not eat corn smut, which they call *ta' chaak*, "excrement of the rain.") However, a strange, serious disease struck the drought-affected cornfields of Presidente Juarez in 2001, killing plants that grew from late replanting efforts. Milpas are scattered, isolated, and frequently relocated; insects have no opportunity to build up substantial populations. Moreover, the number of insect-eating birds and lizards in a Chunhuhub milpa has to be seen to be believed. I have counted six species of flycatchers nesting in one. Cattle egrets, doves, and blackbirds home in on larger insects. It is difficult to imagine how any insects survive.

However, leafcutter ants can devastate an isolated citrus orchard. These big ants, which live in colonies of up to several million, are almost impossible to control. Insect-eating birds do not relish them. Their nests are sometimes dynamited. More often, white paint containing insecticide or lime is painted on the treetrunks.

Whiteflies (*mosca blanca*) do not eat much, but they carry viruses that devastate some crops. Usually, predatory insects control whiteflies. They are a problem when overuse of insecticides has wiped out predatory insects.

(Whiteflies are very resistant to insecticides.) However, small whitefly out-breaks frequently devastate household tomato and chile plantings. The ever-innovative Sosa family has found that aloe vera juice kills whiteflies very effectively. Insecticides include Lannate and Chlordane.[2]

Rats, mice, and bruchid weevils threaten stored grain. Mixing the corn with limestone dust and keeping the ears dry in pole-and-thatch store-houses with elevated floors offers control. Beans, seed corn, and other seeds hang in packages or bunches in the roof shade over the household fire; it protects them both by its flame and by drying and smoking them. This does not always work against agile mice. Problems of storage became acute after Hurricane Roxane, when the wet corn that could be salvaged was extremely susceptible to insects and mold.

Traditionally, families could dust herbal powders among stored prod-ucts to discourage insects. Several herbs kill or repel insects. The dangers of using commercial insecticide for this have become known, and most families now do not use them, though some endanger their families by using Lindane and other dangerous chemicals. One or two families tried fumigants that can be put below the raised floor to liberate a gas up into the ears; it dissipates and allows safe eating of the maize. The fumigants turned out to cost more than they were worth. When all these methods fail to stop weevils, infested ears go to the farm animals.

Increasingly, people are storing the grain in plastic sacks. This offers somewhat more protection from damp and the like (though not from the weevils) and obviates the necessity of building a granary.

Mexican storage practices are probably selecting for super-weevils. They certainly select for super-corn. Seed is chosen from the ears that make it through the winter without being eaten, thus automatically select-ing for the most weevil-resistant genotypes. Seed is taken from the middle part of large, even-rowed ears (if the weevils, and family needs, have left enough to allow a choice). Seed selectors avoid the small kernels at the tip of the ear, and the unevenly placed kernels at the proximal end. (Further valuable detail on farming in this general area is found in Murphy 1990; Peraza Lopez 1986; Terán and Rasmussen 1994; Várguez Pasos 1981.)

Storm activity tapers off in October. Winters are quite cold for a tropical location. North winds can drop temperatures almost to freezing on clear nights. Texans say during "blue northers" that "there is nothing between

Texas and the North Pole except some bob-wire fences." Yucatán is south of Texas, with open ocean between. The northers sometimes pick up moisture and often bring fog or even light rain. At other times they remain "blue"—clearing the sky. In either case, they make corn-bean-squash agriculture impossible in the depth of winter, even as far south as Chunhuhub. In winter, trees can be pruned and ground made ready, but this is really a time for resting and for such nonagricultural work as building houses.

Yields in Chunhuhub are fairly typical for the Yucatán Peninsula but are high compared to the drier northwest. Chunhuhub is favored by timely rain and blessed with some of the best soil in the Peninsula. About 5 percent of the ejido land is the best imaginable: deep, rich, fertile, well-drained alluvium that has washed from the hills into fertile basins. Most of this is now irrigated. Another 25 percent is hilly land, with light loamy soils—rich and productive even on the slopes, but truly exceptional in the small valleys. The rest of the ejido is stony flatwoods covered with thin soil or infertile clay, but the ejido's farming population is not great enough—so far—to force people to cultivate this less fortunate terrain. It is left as forest reserve.

Seeding rates for maize, the crop par excellence, range around 10 to 15 kg per ha. Yields on the irrigated land are 1 to 2 tons per ha—even more, with luck and fertilizer. Yields on the rain-watered hill soils are about 900 kg to a ton per ha, but they can reach 2 tons in the wonderful little valleys; on the other hand, they are only half a ton on the rocky, dry slopes, unless *Dios* and the chaak (the Maya rain gods) are exceptionally cooperative. Sometimes they are, but sometimes they go to the other extreme, as in 1995. Drought is commoner than hurricanes; 1994 was a year of hard drought, and the dry seasons of 1996 and 1997 were also serious. In 1996, one could see small fruits dropping in great quantities from the trees, as hot dry weather continued well into May.

Extreme ranges of typical yields are from 350 kg per ha for *mejen nal* on rocky hill soil to 875 on good, from 500 to 2,000 for nalt'eel and other local varieties under the same circumstances, and from 1,000 to 3,000 or 4,000 for hybrid maize, the latter figures for irrigated and fertilized maize in the permanently cultivated and irrigated land. A planter plans on getting 1,000 kg per ha on hill land and 1,500 to 2,000 on good fertile land, and is disappointed if he gets less. Maize grows better in the rich soils, but is also prone to weeds, flooding, and animal attacks; moreover, the light hill soils are surprisingly fertile. So the final harvests are less different than one might expect. Very bad soils—rocky hills and poor, clayey flatwoods soils—yield far less, about 500 (shelled) at most. Figures much over a ton represent very good luck. Farmers continue to cultivate all types of sites, because every year brings some sort of disaster—affecting the fertile but flood-prone lowlands one year, the dry uplands another. When heavy rain or disease destroys the high-yielding but delicate hybrid maize, the tough native varieties usually survive.

Yields are usually given by farmers in terms of sacks per *k'aan*. A k'aan, or *pisik'aan*, the Mexican-Spanish *mecate*, was originally an area measured by a cord of fixed length. (Both *k'aan* and *mecatl* mean "measuring cord"—in Maya and Nahuatl, respectively). A mecate is now set at 20 x 20 meters, giving 25 mecates to a hectare. Holdings and yields are normally given by reference to mecates. A sack holds about 30 kg of maize in the ear, which shells out to 20 to 25 kg of grain. Some farmers, however, fill their sacks very full, and figure about 35 kg to the sack; others fill only to 25 kg. Thus, the measure is already vague. Normally, people expect at least one sack per mecate, that is, about 750 kg per ha; they hope for two, and they occasionally realize three or four, but only with fertilizing and usually other aids.

At the Classic Maya city of Dzitbanche, south of Chunhuhub in broadly similar terrain, cultivation trials were carried out in the rich black-soil lowlands. Traditional seed and methods were used. Yields ranged from one to as high as three tons per ha (Sergio Delgado Torres, a local farmer who had worked on the project, supplied me with the high figure in 1996; he said it was close to figures for today on rich soils not recently cultivated. The broad, wet lowlands of south-central Quintana Roo are quite rich and almost unexploited. Irrigated rice in the same area yields 6 to 7 tons per ha, and was a major crop for a while, till problems with water delivery and other organizational matters made the rice noncompetitive with imports.)

By contrast, Julián Valdez and I observed the village of San Silverio, which lies in a very dry, stony, semidesert pocket in north-central Quintana Roo, northeast of Tihosuco. Here, maize yields only 500 kg per ha at best, and in both drought years like 1994 and hurricane years like 1995 it yielded almost nothing. Fruit trees (except *abal, Spondias purpurea*) barely hang on in this stony wilderness, and living is harsh. Poverty and malnutrition were the lot of the farmers.

These two cases represent the extremes in Quintana Roo. Chunhuhub and its area lie between.

Hybrid seed, with fertilizer and insecticide, could potentially yield up to 6,000 kg (unshelled) per ha, as it has in experimental plantings by government agronomists, but of course this is thoroughly nontraditional technology. Moreover, it probably does not indicate the potential of even the better areas for maize, for such intensive cultivation on a regular basis normally leads rapidly to grass invasion, making further cultivation difficult and lower yielding. Less rapidly but more insidiously, it leads to soil loss and soil quality loss unless practiced in very fertile, level areas—which yield very well without the extra trouble and expense. Hence such intensive cultivation is very rare.

A twenty-year rotation cycle is necessary in the average plot here. The range is from a year or two to fifty years. This means that an actual hectare of land produces, ideally, 100 kg of shelled maize grain a year: 1,000 kg for two years, then fallow for eighteen. A person needs about a kilo of maize every two days, or 180 kilos per year. Maya may need less, given their small body size and the fact that a household is about half small children; on the other hand, the Maya work terribly hard and have to fight off infections, so 180 per year seems no more than marginally adequate. Certainly, adults I have observed (several hundred persons) and observed closely or interviewed (some twenty) eat about half a kilo of tortillas per day. Others find comparable figures (see Peraza Lopez 1986). This means that it takes about 2 ha of prime land to support a person—allowing a bit of land for other crops and for loss and damage. This assumes that the year is a good one. Assuming average land and a few bad years, we would have to double the allowance to 4 ha.

Thus, a Maya household of six, half of them small children eating half an adult ration, would need 18 ha.

This means that one could support five households (including a couple of larger ones), or, more precisely, about thirty-three people, per square kilometer.

The Yucatán Peninsula is about 120,000 square kilometers, and thus could theoretically support some 4 million people. Actually, much of the peninsula is marsh and dune and other unlikely habitats. The remaining land could easily accommodate 2 million people with slash-and-burn agriculture—given *no* improvements beyond what we see today, and not counting intensive techniques such as raised-bed cultivation, irrigation, and intensive fruit culture.

Even in an awful year, drought and flood are local. Presidente Juarez's maize crop was a near-total loss in 2001, but Chunhuhub had a bumper crop, and even the northeasternmost Presidente Juarez, fields did well. A monstrous hurricane like Gilbert can knock out cultivation over half the peninsula for a year, but people can survive on roots for that long, and such a typhoon is most unlikely to come again in a person's lifetime. Of the three really devastating hurricanes within living memory, Janet hit the center, Gilbert the north, and Roxane the center-south, meaning that no area was directly hit by all three, and few were devastated by two.

Beans range enormously in yield. As noted above, bean-maize inter-crops yield as little as 100 kg per ha of beans. On the other hand, fields in the irrigated land (with its superior soil) have been known to produce 1.5 to 3 tons per ha, but this was with chemical aids—notably, control of pests. Only in a very lucky year will beans yield this high, even with chemistry. The norm is a few hundred kg at most. Part of the difference is a matter of planting time. For high yields they are grown in winter, when pests are fewer and the sun less intense. They are planted in December or January and harvested in the dry season. However, the vast majority of beans are grown in milpas, which means they follow the maize cycle unless planted specially. This is part of the reason for the chronic low yields. Lima beans also yield low; they are harvested well into the winter. Beans brought 5 to 6 pesos per kg in early 1996, but have been rising in price.

Watermelons, one of the most valuable crops for sale, yield 15 to 17 tons per ha but have to be grown in the *mecanizada*, the area irrigated by pumps and plowed by tractors. They need insecticide and fertilizer. There was a local variety called *chay pach* ("chaya-colored back") that was tougher, but the currently grown Charleston grays and other introduced varieties need special expert care.

Animal rearing (see Anderson and Medina Tzuc 2004) was rare in pre-Columbian times. The Maya had dogs, but no other domestic animals until the domestic turkey arrived from the highlands in postclassic times. The Muscovy duck (*Cairina moschata*—a very different bird from the Old World domestic duck) also occurred in pre-Columbian times in the area, but its introduction as a domestic animal will probably never be datable, because wild Muscovies are common in the peninsula (or were common before shotguns). The domestic Muscovy is a South American form, quite different in external appearance from the native one, but probably indistinguishable archaeologically.

With the Spanish came the full panoply of Mediterranean domestic creatures. The Maya quickly named them after their nearest local equivalents. The burro and cow had no local equivalents and became just *burro* and *wakax* (Sp. *vacas*—the plural being borrowed, no doubt, because cattle were always in herds and thus discussed in plural form). Other animals, though, had obvious local relatives. The horse was assimilated to the tapir, *tsiimin*; today the word has become transferred so completely to the horse that the Maya use the Spanish *danta* or *tapir* (both borrowed from other Native American languages!) to refer to the tapir itself. The pig was named after the white-lipped peccary, *k'eek'en*; the wild white-lipped peccary then became the *k'eek'en ij k'aax* ("forest pig"). The brocket deer, *yuk*, gave its name to the goat. The chicken was the *kaaxlan*, "Castilian," later shortened to *kaax*. This name is already attested in the Motul dictionary at the beginning of the seventeenth century. All these animals were soon adopted as farm stock. Sheep were named not for an animal but for a plant: they were *aj taman*, "cotton animal." Today, the old prefix *aj* has been lost, making the animal indistinguishable from the plant, which can be confusing unless the context is clear.

The Maya of Quintana Roo are enthusiastic animal raisers. They love pets, and they are caring and diligent herders. Unlike all too many impoverished rural people, in Mexico and elsewhere, most traditional Maya are gentle and thoughtful with animals. (Chickens are the great exception. Chickens thrive on neglect—and get plenty of it.) Animals may suffer from hunger and disease, but only when the people do. A glance at the dogs is an instant and accurate guide to the economic status of a Maya village. Chunhuhub's are usually not fat, but at least they are usually sleek and happy except after a hurricane. Dogs in early 1996 were very thin, and so were their owners.

Every Maya garden has its dogs, chickens, and turkeys, and larger ones usually have Muscovy ducks (which sell for 12 pesos or more—a good deal of money for a bird) and pigs. INEGI (Instituto Nacional de Estadistica Geografia e Informatica, Mexico's statistics bureau) recorded some 20,000 pigs in the municipio in 1994, and this was certainly a huge underestimate (INEGI 1994:102). The Maya know that pigs eat the garbage and weeds in the village, thus vastly improving its sanitation and health picture. Also, the street potholes make perfect wallows.

Cattle are common on private ranches, but not in the ejidos. Only 5,023 cattle were recorded for the municipio in the latest agricultural census (INEGI 1994:92), and, since cattle raising is generally a highly visible undertaking, this is much less of an undercount than the pig census. In the old days, private ranching carried a risk; one person—not noted for his easy or yielding ways—said that he had a ranch some forty years ago in a remote area, but people kept trying to take over his ranch or cattle at gunpoint, so he gave it up and moved to town. There are, currently, very few people ranching in the Chunhuhub area; I am aware of half a dozen ranches, all but one having fewer than a hundred cattle. A few other people have one or two "scrub" cows on small properties; for them, cows are nothing but a minor way of banking cash, a walking savings account in a diversified agricultural portfolio. Cattle bring about 800 pesos for weanlings, 1,500 (or about 6 pesos per kg on hoof) for cows, 7,000 for stud bulls (2001 figures). Sales are mostly to local butchers and casual buyers. Other nearby communities like Presidente Juarez have a few more, but cattle are not important in this part of Quintana Roo.

An example of "modernization" gone mad is found in *ganaderización*, the repeated attempts to convert Mexico's tropical forest to cattle pasture. This has repeatedly failed in the Yucatán Peninsula and yet is repeatedly reattempted. Small, local successes have occurred when extremely labor-intensive maintenance of pasture is undertaken, but the economics of the enterprise are highly questionable in these cases. In Yaxley (eastern Quintana Roo), the federal government cleared vast areas of land in 1979, but never delivered the promised cattle. More than a million pesos were spent; the land was not only worthless for pasture, but its value for forestry and milpa cutting was lost (Hostettler 1996:187–188). This is a typical story all over Latin America (Painter and Durham 1995).

Where good stock is carefully raised on natural grassland, or sometimes good brushy land, ranching can pay and be a reasonable use of the tropical environment. In the Chunhuhub area, Gonzalo Chan's small ranch (forty Zebu and Indobrasil cattle) and the García family's large one offer examples. Natural savannahs for which cattle are ideal exist around Valle Hermosa, south of Chunhuhub, where one can see true old-style Mexican cowboys who have relocated from traditional cattle country in central Mexico. Many Maya, as well, have learned to be *vaqueros*. In such natural grasslands, even dairying (which requires good feed) succeeds, and farmers make excellent cheese.

However, far more typical was the case of Presidente Juarez. About five square kilometers of seasonally wet terrain were cleared in a 1980s' government scheme so ill-advised as to border on the insane. The original plan was to grow rice, but the soil and hydrology were both unsuited to the crop, which yielded fairly well for three years but then declined while weeds increased. After that, the community attempted to use it for pasture. The land is almost worthless for good pasture grass and promptly regrew to weedy, coarse grasses. The project got mired in politics (some rather shady—as is typical of large-scale land clearing in Mexico and elsewhere). The whole area was abandoned. The vast expanse of dry grass was a natural firetrap; burning, and the tough sod of the weedy grasses, prevented forest regeneration. The land continues as rough pasture, for sheep and cattle, but supports very few of either. There are, moreover, other local problems with stock rearing. In November of 2001 a jaguar ate several sheep. The men of the community banded together and spent a day beating the bush (literally) with guns and dogs. They found not a track. The next night the jaguar returned and ate more sheep.

Another large clearance in forest was made in a nearby ejido around 1991, but proved hopeless for grass. It grew back to native weeds. In April of 1996, it burned, carrying fire into neighboring productive forest. As of 2001 it was slowly returning to healthy second-growth woodland. Chunhuhub had a similar experience, with a happier ending—the forest regrew. After twenty years, it could be cut for fields once more.

Nonnative grasses have made themselves a problem everywhere by their easy combustibility and their habit of carrying fire into good forest. They may slowly and insidiously take over much of Mexico, destroying its natural plants and animals entirely. This has happened in much of Chiapas and Tabasco and, farther afield, in Sonora and Sinaloa. The magnificent tropical forests of the foothills of northern Chiapas were still about half intact as of 1991, but are now all gone or extremely degraded—not due so much to clearing as to wildfires carried by tall nonnative grass (personal observation). Thus even forests on inaccessible slopes have disappeared.

The economics of ranching are such that only well-managed ranches succeed. Poorly designed government schemes cannot be anything but disasters. It is hoped that such madness is a thing of the past in Quintana Roo. Not only in that state, but throughout much of Mexico, ganaderización has been a terrible mistake. Urban bureaucrats, thinking only of increasing the luxury food of other urban well-to-do folk, have sacrificed Mexico's biodiversity, without the slightest economic or nutritional excuse. The process is old; colonial authorities favored livestock over indigenous farming, for reasons that included ethnic prejudice, and this set a pattern (Melville 1994, focusing on sheep and goats; Andrew Mouat, unpublished research; Painter and Durham 1995, focusing on cattle. Melville's conclusions have been challenged, but Mouat's exceedingly detailed field and documentary research in Oaxaca proved that Melville understated rather than overstated the problem). Livestock does very well in Mexico's natural grasslands, and some local cattle varieties flourish in thorn scrub and other rough habitats; here, ranching can flourish and even protect the environment (Kaus 1992). The problem occurs when healthy forest on land poorly adapted to grass is cleared for ranching.

There are some genuine economic reasons for such clearing, however. One point not made in the otherwise excellent work on ganaderización by Painter and Durham (1995) is that cattle provide a return right here and right now, while the other benefits of the rain forest are not adequately compensated. When they manage the forest in the traditional way, the Maya are protecting carbon fixing capacity, biodiversity, medicinal plants, North American migrant birds, local rainfall, local soils, and other goods of local or world importance. They get not one cent for any of this. The combined value of those services is certainly very much greater than the value of a few scrub cattle. Yet the Maya get the money for the cattle. (These are sold locally. At least, hamburger merchants do not get the profits here.) So the forest goes and the cattle stay.

Another reason for ganaderización is that Maya today must usually work at town jobs, as well as rural ones, to support themselves. They do not have time to be full-scale milpa cultivators. Cattle take very little work.

Already, in 1987, a reported 39,201 sq km of Quintana Roo were affected by extreme soil and environmental degradation due to this and related causes; degradation was severe on 23,521 of these (INEGI 2000d:91; this seems to be the most up-to-date figure). This was underreported, and the situation has grown worse since. I would estimate the current area to be at least ten times that high. Soon, if nothing is done, Quintana Roo and all the New World tropics will be a sea of introduced weedy grasses.[3]

But to return to our muttons—literally. Sheep are kept by a very few people (there were only about three thousand sheep in Felipe Carrillo Puerto municipio—INEGI 1994:117). One shepherd, Carlos Perera Ake, has for many years kept a flock of around fifteen. A gentle, hermitlike soul who loves the company of his sheep more than that of most humans, he raises them for meat, selling them alive for 5 pesos per kg. Wool is worth 30 pesos per kg, but these tropical sheep bear virtually none. Maya do not normally eat lamb or mutton, so the sheep go to Cancun and Chetumal, where non-Maya consume them. As of 2001–2002, sheep were more widespread, and keeping sheep was becoming a regular thing to do. Eating sheep had become regular, if not very common. They are cut up and pit-barbecued in the earth oven (*pib*).

The most hopeful development in animal rearing is the raising of locally endangered game animals. I enthusiastically propagandized for this (losing all claim to ethnographic objectivity), especially after seeing a successful peccary farm in San Francisco Ake, an isolated town north of Chunhuhub. Felix Medina Tzuc tried to find a mate for his pet peccary, Bixa, but he failed, and she took matters into her own hooves by running off to find one in the forest, where, unfortunately, she was shot by a hunter. I was able to encourage another of my field helpers, Andrés Sosa, to raise two pacas and try to breed them. However, this too failed. Paca meat is the absolute favorite of the Chunhuhub Maya, with the sad result that these charming animals are almost exterminated from the area. (Pacas eat fruit, and are famous enemies of squash; the related agoutis prefer roots.) More hopeful are the efforts of the University of Yucatán and of some private ranchers. Raising of deer, peccaries, and pacas is still experimental but works well and might easily become general in the near future.

The new world of tractors and orchards has made little change in this age-old order. Less than 5 percent of Chunhuhub's land is flat, extensive, well-watered, fertile valley soil, suitable for irrigation and tractor cultivation. This *riego* (irrigation area) was developed in the late 1980s.

The government supplied a tractor, but it soon broke, overwhelmed by trying to cope with the rampant Johnson grass that had unfortunately been introduced with the irrigation project. It remained broken for five years, but was running again from 1996 on. As of 1996, it cost 200 pesos per ha for a quick run over the land (*barbecho*) and 500 for a full treatment (*rastrejo, barbecho, surcando*—dragging, basic plowing to dig the weeds in, and a second furrowing for seeding).

Most of this land produced watermelons, chiles, tomatoes, and maize. A few small orchards flourished. Interest in the whole project was luke-warm. It produces more than the milpas, but it involves more problems with weeds and is culturally uncongenial. Most families prefer their scattered milpas.

The orchards are primarily in the form of *corredores citricos*—"citrus corridors." In the 1980s and early 1990s, belts of land 1 or 2 ha deep were cleared along major roads, equipped with tube-wells and pumps, and opened up for citrus and other tree crops. At the time, government and Maya alike planned to develop orange juice concentrate as a major export from the peninsula. This failed, in spite of a fairly successful plant at Akil in Yucatán state and later a small plant in Chetumal.

Yucatán's long dry period, unreliable rain, and thin, rocky soil, and the lack of a specialized, skilled labor supply, guaranteed that the peninsula cannot compete with Florida or Brazil. Moreover, the tube-well pumps are constantly breaking down, and spare parts are difficult to come by. The oranges survive and bear without irrigation, but do not produce well—less than half as well as they would with reliable water and good soil. Fertilizer, at around 120 pesos per bag, is expensive, but superphosphate can make a large difference, especially on chak lu'um.

However, there is a bright side. The harsh conditions and thin but fertile soil produce citrus of exceptional quality. Local growers maintain that "experts" have told them that the juice is the finest in the world. It certainly is exceptionally good, according to all who have tasted it.

Thanks to this high quality, the explosive growth of tourism in Cancun and other tourist sites created a major new outlet for citrus. Tourists, restaurants, and local workers agreed on one thing: they all wanted all the fresh local fruit they could get. Any fresh fruit commanded a ready market at a high price. Creative farmers enthusiastically intercrop such things as tomatoes, chiles, and root crops with the citrus, thus making the whole operation more productive and culturally congenial. Pruning is limited, but one must prune off *chupones* (Sp., "suckers," including water sprouts) from the base, though they are encouraged when coming from major branches.

The original government plan was strictly limited to juice concentrate, but growers quickly learned that urban consumers wanted fresh fruit and juice and would pay premium prices for these commodities. In Mérida and Cancun, grower organizations began marketing first oranges and later the juice also on a large scale, selling it on street corners and in small shops, in plastic bottles. Finally, the idea of marketing fresh orange juice spread to Chunhuhub, where food stalls in the market now sell it. At least one lunch stand has added pineapple, watermelon, and other *licuadas* (blended fruit pulp, with purified water).

Oranges yield about 5 kg of fruit per tree in their fourth year, meaning about 1,615 kg per ha at a standard stocking rate of 323 trees per ha. Well-cared-for trees yield 40 kg in their fifth year, then on up to stable yields of around 140 kg at ten years (45,200 kg per ha). More typical Chunhuhub trees yield such amounts only if planted in very fertile soil that holds moisture well, and then only if the year brings neither drought nor flood. Once, fruit was measured in *almuds* (an old Moorish measure, later standardized at three and a half kg), and some boxes of this size are still found.

Careful orchard-keepers paint the trunks with pesticides such as Bordeaux mixture—copper sulfate, made up here with lime and alum and water. Other pesticides are advised but rarely used. Leafcutter ants are the only serious pest (see above). Fruit flies (*Anastrepha* spp.) are a minor problem.

Andrés Sosa calculates that a full state-of-the-art citrus orchard would cost 116,285 pesos to install—this includes preparing the land, buying the trees, putting in irrigation, and so on—and would yield 132,042 pesos per year when the trees came into full bearing, with costs around 14,000 pesos per year (forty-six man-days involved; these are 1996 figures but have not changed much as of 2001). This would give a favorable return even at high interest rates. Of course, this is the balance for a fully industrial orchard, with heavy use of fertilizers and pesticides. The typical Chunhuhub orchard costs a tiny fraction as much to put in, and almost nothing to operate, but it returns correspondingly less.

The fruit can go for juice concentrate, but prices are low—270 pesos per ton, about four thousand fruits—and often there is such a glut that no one buys at all. Supposedly, the juice is sweeter and more flavorful than Brazil's, but Brazil is a much lower cost producer. Most of the juice goes to the United States, and export levels have fluctuated wildly (sinking to nugatory levels in 1985, and usually remaining there). So people sell individual fruit to the urban market. In 1996, fruit sold, from the farm gate, for 20 to 80 pesos per hundred, for the fresh-fruit market in Cancun. The spread reflects, once again, Hurricane Roxane; before it, a bumper crop had dropped the price to 20 pesos; after it, the few oranges that could be found sold for 80. Limes went from 5 to 50 pesos per box in the same period. By 2001 prices had sharply increased to around 100 pesos per hundred. The crash in tourism following the terrorist bombings in September of that year led to a decline to 70 to 80 pesos at the farm gate, a price ruinous to small producers.

Other fruits are not of much commercial importance. They are grown on too small a scale and bear too erratically. I observed three mamey trees, near each other and each 10 meters tall, that had about a hundred, fifty, and five fruits on them respectively. This enormous difference was a function of age, soil, and care; the five-fruit tree was young, planted in a rocky place, and given no care. A spindly, shaded anona tree 5 m tall had only four fruits, a miserable yield; other, healthier trees of that size may have fifty.

Mango trees yield up to five hundred fruits per tree per crop, with a heavy harvest in spring or early summer; they also bear fruit in small amounts throughout the year. Achiote is important, bringing 20 pesos per kg or more for shelled seeds. Some people have small achiote orchards, but typically achiote is grown in dooryard gardens. A tree bears only a few kg per year. Minor crops include coffee and cacao, but these grow only for home or local use. Cacao does not grow well enough to allow the successful commercial development found among Maya communities in Belize (Emch 2003) as well as in larger-scale commercial agriculture in Chiapas, Oaxaca, and Tabasco in Mexico.

The effect of Roxane on fruit and vegetables was even more serious than on maize. The fruit crop was blown right off the trees. Almost everything was a total loss. Even the normally indestructible chaya, which was a traditional disaster food in precisely such circumstances, was hit hard. The experience was productive of much insight as to why agriculture remains simple and remains focused on hardy rather than high-yielding crops. More generally, hurricanes and droughts (such as those of 2001) constantly remind the Maya and their observers that subsistence agriculture is no way to make a good living unless one works terribly hard. A constant drain of young people to cities is a result.

Vegetable raising is rare but very profitable. Chiles brought up to 15 pesos per kg in 1996 and substantially more in 2001. The García family grows chiles commercially, planting 26,000 plants per ha, and getting 200 to 300 huacales (boxes of ca. 10 kg) from this. But they use modern agricultural techniques, with plastic shelters and pesticides.

I brought presents of American vegetable seeds, and my friends soon learned that American garden vegetable varieties are not as tough as local ones under local conditions. Pigs, peccaries, whiteflies, and virus had a field day with the shoots, and even the briefest drought killed them. The difference was particularly striking in chile peppers; American varieties were too delicate to survive, while local strains are noted for their hardiness. Even so, the American seeds were greatly appreciated, simply because they were something new to try.

Mameys, grapefruit, limes, papayas, bananas, anonas, avocados, and other fruit are sold and have enormous potential for development.

Nonfood uses of the vegetation are slowly fading out.[4] The complex, sophisticated knowledge of how to build a pole-and-thatch hut is still alive, but masonry houses are replacing the old *bajareques*.

Agricultural costs and wages are low. As of 1996, it cost 300 pesos to get a hectare of milpa cleared, if one hired workers. For 75 it was sowed, and 125 got it weeded—basically using machetes to chop back the brush and herbs; clean weeding, grubbing, and pulling up everything took more money. A typical field needs two weedings in the growing season. For 150 pesos you could get someone with a truck to take your harvest to town. A skilled worker on a commercial farm got 800 to 1,250 pesos a month; unskilled workers are hired casually, by the job, and were lucky to get 20 pesos a day. Cowboys, for instance, work at that rate. By 2001 the figures had sharply increased, casual work paying 30 to 50 pesos per day (depending on the nature of the work, and on the supply-demand structure of the local labor market).

This may be compared to the shadow "income" of cashless cultivators in the forest. A truly traditional Maya cultivator, living in a virtually cashless subsistence economy, produces goods and services that would cost (in the late 1990s and early twenty-first century) 14,000 to 20,000 pesos per year on the open market. That is, he or she is, in effect, earning about 1500 pesos per month. The figure would once have been higher, since traditional subsistence items like game and valuable woods were previously available in much greater amounts.

In 1991, SARH (Secretaría de Agricultura y Recursos Hidraulicos), Mexico's department of agriculture, had an office at Tampac, a ranch 5 km south of Chunhuhub. It theoretically oversaw aid and loan programs and collected statistics. In fact, the office was very rarely occupied. When I inquired about agricultural statistics there and later in Felipe Carrillo Puerto and Chetumal, I found there were none. Collection of such data had been abandoned. The SARH office at Tampac was closed, permanently, around 1995—a loss not felt by anyone, so far as I could discern. It was simply too difficult to count the thousands of acres of maize, squash, beans, watermelons, chiles, and everything else that the Maya grew for subsistence and local sale. It is easy see why statistics are not collected, when one tries to do it oneself! Maya plots are small and scattered. Harvest is not all at once; people use the field as a storehouse, and take a sack of maize or squash as they need it. This leads to major losses to animals, but one can always eat the animals. Statisticians throw up their hands; how does one count maize harvested a few ears at a time over six months, and frequently cycled through game animals on its way to the human consumer?

Article 27 of the Mexican Constitution, allowing for the privatization of ejidos, led to the ending of federal ejido programs. Moreover, banks had been privatized, and the special agricultural and ejido banks—especially Banrural, the public bank for agricultural development—no longer made loans; farmers had to go to the private sector.

In 1991, labor, administration, preparation, fertilizer, seed, irrigation, insecticide, and a bit for insurance were all covered and explicitly spelled out in the Banrural loan packages. Formerly, the bank actually provided the fertilizer and insecticide. By 1991, Banrural had dropped this, but demanded accounting of expenditures on the above. However, this permitted flexibility in what and how much fertilizer and pesticide to apply. SARH provided recommendations, which ran to large amounts of a wide variety of pesticides.

Every ill wind blows some good, and the decline and fall of public agricultural banks at least freed the countryfolk from "package" loans. One of the initially brightest ideas of the green revolution was to make loans for specific purposes: buying hybrid seed and the pesticides and fertilizers necessary to make them grow well. Unfortunately, chemical companies lobbied industriously (and, according to persistent rumors, paid certain officials under the table) to have Mexican packages that included far too much pesticide and fertilizer. Exceedingly dangerous levels of these chemicals were forced on rural people. In many parts of Mexico, the countryfolk were unaware of the dangers of these chemicals, and there were serious problems (Wright 1990). The Maya were more alert, and most of them used the chemicals very carefully, or simply avoided the whole matter by not taking out package loans. The packages were abolished in the late 1980s, and the whole rural loan program died in 1993.

With no bank in Chunhuhub and little chance of a poor family getting a loan from any private bank, the chances for credit ended, as far as most of the population is concerned. Orchard growers and ranchers on large parcels deal successfully with the private sector. Still, they have to pay high interest. Figures of 37 percent and more are recorded. One person borrowed 18,000 pesos in 1995 and had to pay back 30,000 in a year, after Hurricane Roxane. He had a business as well as a farm and could repay, but few other farmers could repay their debts that year. More typical rates are around 7 percent interest per year, still enough to deter less-than-affluent borrowers.

In a normal year, people almost always repay, but often late. However, Roxane led to massive defaulting by ruined farmers. Crops had been insured through the government's rural bank until a few years before, but now insurance, like capital, is expensive.

Because of the intensely familial nature of Maya society, there are no informal community savings and loan associations, as there are elsewhere in Mexico (and the rest of the world). Without credit or a well-to-do family, no one can buy hybrid seed, irrigation pumps go unfixed, and highly desirable but slow-to-mature experiments such as mamey orchards and timber plantations rarely get beyond the contemplation stage.

A rural advancement agency, Procampo, will advance 480 pesos per ha outright, and Pronasol (a political program associated with the PRI party) loans 400 pesos per ha (1996 figures). If one repays this in a year, one can potentially get 800 pesos the next year.

Chunhuhub has a small ejido nursery started with government support, and a larger nursery that is strictly a government agricultural operation. The former produces valuable timber trees for reforestation, and fruit and ornamental plants for gardens. The latter is strictly a citrus operation (though it has experimented with things like cinnamon).

The troubles and prospects of farming and the environment in Chunhuhub are best summed up in a story—an exceptional one, but all the more revealing because of its exceptional nature.

In 1991, Andrés Sosa was a slender young Maya man whose farm bridged millennia. Around 1972 his father, Serafino Sosa Castillo, moved from Campeche to the wild frontier of Quintana Roo. They settled in Chunhuhub, then a small village recently carved out of the tall, ancient forest that still covers much of that state. Here they were given free use on leasehold of a small parcel of land. The ejido had plenty of land to spare; immigrants, though lacking ejido rights, were welcomed.

Andrés, a young child at the time, grew up close to the land, and eventually went to CBTA-80, Chunhuhub's technical training school—equivalent to a vocational junior college in the United States. In a farm town like Chunhuhub, the "CEBETA" naturally focuses on agriculture. (As we have seen, it has also added computer training and has several modern microcomputers—a fascinating sight in a town that, when Andrés went through CEBETA, lacked telephone service or mail delivery and lives by ancient Maya slash-and-burn agriculture.) Andrés studied agriculture there. Unlike most of his classmates, he stayed home to use his knowledge. The usual progression of CEBETA graduates is to the city, where their training allows them to become urban workers. Often, they become low-level rural workers for SARH and spend their lives collecting crop statistics or shuffling forms—using none of the high-line agricultural knowledge they acquired. This is sadly typical of education throughout the world. The "certification" function dominates, and the vast amounts of effort invested in teaching and learning actual content is simply thrown away.

Andrés was a different man, and his family was different. They were dedicated farmers, and stubborn optimists. They felt that, with hard work, farming could pay and be a more satisfying life than that of the cities. Andrés saw no reason to join the depressing brain drain that has depleted Mexico's—and the world's—rural areas without deeply benefiting the cities. So he stayed on the farm with his father and two brothers, to try to combine the new knowledge taught by CEBETA with the vast store of wisdom accumulated through many millennia by the Maya of the Yucatán Peninsula.

In 1987, the Sosa family acquired a 2-hectare parcel of ejido land on the north edge of Chunhuhub. This they developed into a multicrop farm. A dirt road branched off the main highway, traversed the typical flat limestone of the Peninsula, passed through a few yards of thorny second-growth forest, and entered the farm. Young trees rose along it, shading the way.

The land was typical of Chunhuhub: rolling limestone with a thin covering of red soil. This soil—*chak k'aankab* ("red-yellow soil") in Maya—is more fertile than it looks, but it is still shallow and rather nutrient poor. On the flattest and best parts of the Sosa farm, it had more organic content, more clay, and more moisture—verging on 'eek' lu'um (black land). On the steepest and worst parts were outcrops of tsek'el (thin lithosol). The natural cover was tall, drought-deciduous forest. Fairly tall (to 15 m or more) second growth persisted on the northwest, providing a corridor for animal pests such as raccoons and coatis to enter the fields. On the other sides, *juubche'* (thorny second growth) ranges from 2 to almost 10 meters in height. Many plants of the second growth are legumes and have nitrogen-fixing bacteria in their roots. Thus they enrich the soil. However, nutrient deficiencies (especially of metal ions) in the limestone soil limit the growth of these bacteria, as of much else.

The Sosa farm possessed some good soil, reserved for the high-value crops. The less good soil was reserved for tree crops. Still worse land on the stony slopes was given over to milpa corn agriculture—Maya corn can grow almost anywhere. The worst soil was fenced to raise sheep; in 1990–91 about thirty were raised. They were sold on the Cancun meat market when the winter-spring drought dried up the pasture.

The trees and perennial crops occupy the southeast part of the parcel. These included achiote, coconut, lime, mandarin orange, sweet orange, grapefruit, mango, chaya, almendro (for shade; not used for food), papaya, pineapple, banana, tamarind, guava, breadfruit, ciricote, guanabana, avocado, cashew, nance, xa'an (*Sabal* sp.), sapote (including a large old wild one left when clearing), cacao, coffee, mamey. All these bore fruit. Few of them were large enough and productive enough to provide much surplus for sale, and the cacao and coffee were frankly experimental, since they do not grow well in Chunhuhub. Papayas, bananas, and citrus—all fast-growing—reached a size that allowed production for sale. The mameys and avocados were 7 or 8 meters tall, but not yet bearing heavily. A few ornamentals were planted.

Among the trees grew many cucumbers, chayotes, and squashes, including an obscure variety of pumpkin called "tabasqueña" (apparently a form of *C. pepo*). Bottle gourds included the small-fruited ones called *tuch* in Maya. These cover the ground and keep weeds down, while not competing with the tree crops. A few small perennials were grown: yams (*Dioscorea*

Andrés Sosa tending squash in his garden.

sp.), macal, three aloe vera plants for medicine, some sugar cane, and some Taiwan grass, mainly just to have it, but potentially for sheep and cattle feed. The local "oregano," *Lippia dulcis*, was in evidence.

In the better soil, centrally located in the parcel, the main crop was watermelon of the variety "Charleston gray." Many vegetables were also grown, in rotation with or in different places from the melons. The vegetables were grown in raised beds. These were on the Chinese/French model, about 10 to 20 cm high, 1 to 2 meters across, and 10 or more meters long. They were separated by drainage/irrigation trenches. Seedlings were raised in special seedling beds and transplanted to the bigger ones in the field. Tomatoes were the most important. Most of these were large ones, but little semiwild cherry tomatoes (*tomates de costa*, because they are said to grow wild on the coast) were grown. Various kinds of chiles were grown: sweet (bell), xkatik, *poblano*, *guajillo*, and *verde* (all *C. annuum*); Tabasco pepper; and the Maya favorite, the habanero. The maax is a short-lived perennial, and thus grew among the fruit trees rather than in the fields. Kohlrabi, beets, lettuce, cabbage, cilantro, carrots, and other vegetables grew as season and space could afford. So did *jamaica* (*Hibiscus sabdariffa*), grown as an annual. Its flowers make a sweet-sour herbal drink. Beans included ordinary black beans, cowpeas, lima beans, and pigeon peas. A small wild bean was strictly a green manure. Beans need phosphorus and potassium here, and thus were fertilized. With this, they yielded up to 1.5 tons per ha, an exceedingly high rate for the area. Many were, however, picked before they matured enough to yield such weights, and sold as green beans (*ejotes*). Seeds were bought, typically in the nearest large town, Jose María Morelos.

A few hives of European bees flourished in an isolated corner. Africanized bees have taken over in most of Chunhuhub, but the Sosas carefully rogued out Africanized queens.

The Sosa land occupies a small depression with a perched water table, a rather common and highly prized type of environment in Chunhuhub. Water is only 8 meters down—as opposed to more than 100 meters (often 150) in places without a perched aquifer. The Sosas used drip irrigation when possible, with flexible, movable pipes. They had a well with an electric pump and a manual backup. It fed a small masonry tank, from which siphons led to the irrigation pipes. A tractor filled out the list of farm machinery, but handwork was more important.

Maize yielded 1.5 tons per hectare, or, with incredibly good luck (especially in regard to rain), almost 2. Holes were made one double-pace apart—about a meter or a bit more—with a digging stick. Four kernels were dropped in each hole, for a seeding rate of about a kilogram per mecate.

The Sosas grew not only local but also hybrid corn, and had to buy seed for this, as well as seed for *maiz palomitero* (popcorn). Preparation of the cornfield was by the inevitable slash-and-burn system. Fertilizer added little to the mix of limestone soil and ash. The limits on maize growth in this part of Chunhuhub are usually sct not by nutrients, but by the shallowness of the soil and by any lack of rain that occurs in the growing season.

Over the years, the Sosas invested 100 million old pesos (100,000 new ones), not counting the 50 million per year costs for seed, fertilizer, help with clearing and weeding, and so on. The Sosas did not ask for or receive credit, and were rarely if ever in debt. Profits reached 70 million old pesos gross—20 million net—in a year. Up to 20 million or more came from watermelons; a 10-mecate plot could produce 10 tons of these, bringing (gross profit) 40 million. Two crops per year were possible. A crop in winter would bring less than the above, but one in late spring or early summer could bring substantially more, since watermelons were scarce then, and the savage dry heat made watermelons desirable to urban buyers. Cancun, in particular, presents an inexhaustible market for Chunhuhub watermelons.

Tomatoes sown in early spring yielded 20 boxes—half a ton—by early summer, but then tended to succumb to virus carried by whiteflies, a problem that came to the Yucatán around 1990. The Sosas were aware of biological control (used effectively against the whitefly in the United States), but had no way of getting information on that. The pesticide companies saw to it that SARH had plenty of information to hand out on the virtues of chemistry, but no one stepped forward to do the same for biocontrol.

Mammals such as gophers, coatis, raccoons, and deer, and birds such as jays and blackbirds, were highly visible but did trivial damage. Diseases included *tizón*—"firebrand," a fungal wilt that looks as if it burned the plants. Insect pests included *pulgón* (cucumber beetle), *conchuelita* (a flea or shot-hole beetle), and several other "bichos," as well as the *minador* (leaf miner), *usito* (leafhopper; a "mestizo Maya" word, combining the Maya for gnat, *uus*, with a Spanish ending), and others. Malathion, copper fungicides, Lannate, and other pesticides controlled these. The Sosas recognized the dangers, and tried to keep pesticides down and use every possible strategy to control pests by less toxic means. They knew that birds and insects ate a great deal of the pest population. Companion plantings were used to discourage pests. They experimented with a modern range of chemical fertilizers; besides using side-dressings for maize and other crops, they tried foliar fertilizers and growth hormones. However, interplanting of beans and peanuts with other crops, and use of beans for green manure, supplemented the artificial fertilizers. Everything possible was composted, and the compost worked into soil or spread around plants. (This presents a potential for infection with pests, however, unless it is carefully done.)

Watermelons were the main income source, followed by tomatoes and other vegetables; tree crops were insignificant, and the maize was basically for personal use. The Sosas did not eat much meat; most of what they ate was chicken of their own raising. They also got some from the mammalian pests that occasionally invaded the garden. A paca could be lured by whistling into a folded leaf, and then one shotgun shell provided several kg of good meat. At their house, a short bicycle ride from the farm, they raised a few turkeys and pigs as well as chickens.

Thus, the Sosas developed a highly successful small farm in an area usually considered extremely bad for agriculture.

Obviously, this record has implications for the region. The clearest implication is: diversify. The worst bane of the peninsula has been the search for the single quick fix. Monocropping is bad almost everywhere in the world, but nowhere more so than in the Yucatán Peninsula (Faust 1998; Gates 1992, with important data; Kathleen Truman, personal communication 1991). Poor soils, a climate conducive to pests, isolation, and relatively apathetic investors have generally guaranteed that the peninsula was the first to suffer all the ills of monocropping, and often the last to reap any benefits. The henequen growers lost out to foreign competitors even before nylon wrecked the henequen market for good. Orange orchards are not competitive with those of Veracruz or Brazil. Intensive tomato and chile growing at Dzidantun has succumbed to the whitefly. By contrast, the Sosas grow a wide variety of things, and thus always have at least some of whatever is commanding a good price. They do not invest in one thing, only to see its price crash as everyone else climbs aboard. Nor are their fields wiped out instantly by explosive pest increases.

The next most obvious lesson is: stay small. Stick to a farm that can be worked by traditional methods, supplemented (not replaced) by modern ones. Do not sacrifice the wisdom of millennia for the current fads of agribusiness, and do not forget that there is no substitute for careful personal effort. Above all, do not invest heavily in purchased inputs; this would run up debts without necessarily improving profits. The tendency among farmers is to decrease labor input when capital input rises; this leads to an insidious erosion of the farm if monitoring and personal attention are sacrificed.

Less obvious, but perhaps in the long run more significant, is this lesson: maximize knowledge. The Sosas were unique (for Chunhuhub) in their devotion to learn both traditional Maya lore and modern CEBETA lore. Better still: they kept their ears open for any new information that could put them ahead.

On the other hand, they were only slightly ahead of others. Many other families in Chunhuhub practice some or most of the Sosas' intensive strategies and tactics. No other family used so many, but, on the other hand, nothing the Sosas did was really unique in Chunhuhub. Other families were more prone to follow the strategy of household diversification into shopkeeping or skilled trades. The other family that practiced large-scale raised-bed farming, for instance, included a teacher, a newspaper dealer, a barber, and two or three men who worked in construction when not raising vegetables. Their formidable talent for learning new ways of survival has thus been applied mainly to domains outside of agriculture.

The archaeological Mayanist may be more interested in what the Sosa farm has to say about the past. Strategies that were almost certainly continuous from ancient Maya times included highly diversified multicropping, heavy investment in tree crops, selective and strategic use of companion plantings, leguminous cover crops, and compost fertilizer.

The ridged fields in wetlands of Quintana Roo and other Maya lands imply an ancient fondness for land rearranging that must surely have expressed itself in raised-field systems like the Sosas'. Though the Sosa system was influenced by teaching, small raised-bed systems are widely used in Chunhuhub by people who have no exposure to modern agricultural technology and who appear to be preserving a very ancient custom.

I am also convinced—following suggestions from a former student, Sharon Burton—that the Classic-period Maya must have had higher-yield but more delicate and care-requiring crops than are historically documented for Yucatec farms. The Sosas' careful attention to varieties, seed selection, and seed storage seems to me to hark back to ancient ancestors.

Indeed, the Sosa farm seems to me an excellent working model of the farms that fed Tikal, Uxmal, or Coba, except for the rather minor and quite dispensable use of chemicals and machinery. I think that the ancient farms would have had greater skills at pest control—perhaps through companion plantings or encouragement of predators.

I think—and I like to think—that the ancient Maya were very much like the Sosas: hopeful, hard-working, patient, pragmatic, and quick to seize on any new bit of information that promised to be of any use to them. Perhaps the kings were involved with war and sacrifice, but I suspect that even the kings spent much of their time organizing agriculture. As for the vast voiceless majority—the millions of Maya, over thousands of years, who are known to us only from modest, silent house mounds in the forest—I think the Sosas give them an authentic voice.

When I returned in 1996, the Sosa farm was a pale reflection of its former self. The family had followed most other entrepreneurial families, moving away from a farming focus. As sons moved away and Don Serafino aged, farming became difficult. Moreover, the farm was not paying well. The biggest real problem faced by the Sosa family was marketing. No one except visiting ethnographers liked beets and kohlrabi enough to buy them. Two disastrous years, 1994 and 1995, had damaged crops and devastated local buying power. Tree crops were a glut on the market. Corn for subsistence and watermelons for sale were the only things that looked worthy of much effort in 1996. The family opened a bar in town and made their money selling beer to Maya workers. The farm continued to flourish, but only as a somewhat neglected sideline. It was often worked by men whose pay instantly flowed back into the Sosa coffers, via the beer bar; farm labor in Quintana Roo is thirsty work.

Andrés had become a high-school biology teacher in the urban coastal corridor of the state. Highly intelligent, upwardly mobile, and thoroughly devoted to balanced and sustainable growth, he was studying agriculture, economics, and conservation. He continued to concern himself with agricultural production.

In 2001, the bar was closed; old Don Serafino Sosa had retired. Andrés continues as a biology teacher and lives in Chunhuhub with his own young family. He has become realistic—even a bit cynical—about the future of conservation and agriculture in Quintana Roo.

The Sosa farm showed what could be done by integrating traditional and modern knowledge. Knowing traditional pest-control techniques, such as growing squash vines to cover the ground, allowed them to cut pesticide use by about 99 percent (compared with standard use rates on non-Maya farms). Knowing the value of traditional multicropping allowed them to plan a farm based on intensive but highly diversified cropping. Knowing modern production techniques, and having access to modern seeds and varieties, allowed the Sosas to pick and choose what new ideas to adopt, in order to produce an optimal blend of old and new. They were stunningly successful.

Unfortunately for farming, though relatively fortunately for the Sosas, changes in the family labor force were accompanied by the opening of more lucrative and less effortful ways of maximizing income.

The final message of the Sosa farm is therefore a troubling one. If farming is so uninviting that all the most successful and entrepreneurial families move out of it, what is the future of food production in Mexico and elsewhere?

However, on the basis of the experiences of the Sosas and several other forward-looking and experiment-loving families of Chunhuhub, one could lay out a general plan for sustainable development of the area.

Future intensive cultivation would be most successful if it consisted of small tracts of raised fields, growing mixed vegetables, separated and bounded by windbreaks or groves of various species of fruit trees.

This would, of course, combine with the milpa outfield system, which is perfectly adapted to the lighter and rockier soils of the area—soils that cannot conceivably be productive any other way. Continuous cropping, even of fruit trees, is not viable on these soils. The milpa cycle should ideally be long enough to allow growth of valuable timber, medicinal plants, and saleable ornamental plants between croppings—that is, about twenty to forty years. This would probably maximize long-run returns to land, not only because of the value of the timber, but also because it would conserve the fertility of the soil.

Ecotourism can be vastly increased in the area. Wistful plans for resorts at Rancho El Corozo and elsewhere could perfectly well be practicable, if there were any source of reasonably cheap capital. Meanwhile, other relatively environment-friendly activities, such as hunting, can continue—but hunting will have to be subject to far stricter control than it is at present. In Chunhuhub, all hunting (except of milpa-pest rabbits and coatis) must be outlawed for the foreseeable future, to allow animal populations to recover.

In my youth, in the farm towns of the Midwest and South, I sometimes heard a sardonic line: "Around here, anyone with any git-up-and-go has got up and gone."

Chunhuhub has not suffered from this phenomenon as badly as have the towns nearer to Mérida and Cancun, let alone the villages of Oaxaca and Zacatecas. But it has suffered. Farming still pays, and could pay very much more, but, for many of the young people, the attraction of an assured salary at an indoor job is too much to resist. Chunhuhub's harvest was bad for three years in a row in the early 1990s, at a time when Mexico's agricultural economy was in a state of depression. Meanwhile, Mexican maize is less and less competitive with heavily subsidized United States corn. United States farmers are subsidized by direct payments averaging $55,000 per farm, and indirect benefits (good roads, secure banks, agricultural research . . .) worth perhaps as much again. By contrast, south Mexico's small maize farmers are at least as apt to be shot by the government (at least in Chiapas and Guerrero) as subsidized by it. The North American Free Trade Agreement has led to vast dumping of the heavily subsidized corn on the Mexican market. Small-scale maize farming cannot possibly pay (see, e.g., Rosenberg 2003).

The only reasonable course is to educate one's children and send them out into the world—perhaps keeping one or two at home to work a reduced and more specialized farm.

More typical than the Sosa story is that of the Gongora family. In 1991 they planted 3 ha on the ruins of Tampac—Classic Maya sherds and lithics almost paved parts of their cornfield.

In 1991 the Gongora family was one of the most traditional in Chunhuhub, with one of the best milpas and dooryard gardens. They had pet chachalacas and doves, and were self-sufficient in food and medicine raising many herbs. In spite of being poor and traditional, they were highly education conscious, and the sons of the family were rising in the world.

In 1996 the family was scattered. The mother of the family, a warm, loving, and highly intelligent woman, had died of cancer. The sons had gone to Cancun.

I was fortunate enough to meet my old friend Ariel, one of the sons, on a rare visit. He was doing well, learning English, rising in the world. He and his brothers rarely could get home. We went to see his father. We found old Don Carlos living alone in his traditional bajareque house. The huge dooryard garden was still rich, but he was not able to take much care of it any more. Trees uprooted by Hurricane Roxane lay on the rocks. Don Carlos greeted me warmly, but soon retired into a reverie. We left him alone with his memories.

CHAPTER FOUR

Logging

UNLIKE AGRICULTURE, lumbering is not a traditional Maya industry. One or another type of logging has gone on for centuries in the area, but large-scale commercial logging is a child of the last century. It is, therefore, a far less defined area; no traditional rules govern it, no ceremonies construct an emotional and spiritual side to it. It is, therefore, the source of most of the management conflicts and controversies in Chunhuhub, as in much of the interior of the Yucatán Peninsula.

Quintana Roo has been a major forestry state throughout the twentieth century. It is a major source of forest products, including most of Mexico's mahogany.

In 1994 the state produced 47,506 cubic meters of quality timber: 10,013 of precious woods (mahogany and cedro) and 37,493 of other species. The value of all this was 16.704 million pesos—6.903 million for the precious woods and 9.801 million for the rest (INEGI 1995:205–6). In 1999, the state produced 34,175 cubic meters of commercial timber; 21,107 of this was in Felipe Carrillo Puerto municipio (essentially all the rest was in the southwest, continuous with the Chunhuhub area). Of the municipio's cut, 6,354 was mahogany and cedro, the "precious" woods (INEGI 2000a:225). Especially important was mahogany from Nohbec, Petcacab, Naranjal Poniente, and a few other ejidos that have carefully husbanded their resources. Of course, this reflects only the legal cut. Half of Mexico's timber cut is illegal (INEGI 2000d:2:585). However, in Quintana Roo (but not, alas, in northern Mexico), much of this represents local small-scale cutting for houses and the like.

To this may be added the production of chicle, worth 3.601 million pesos in 1994, and palm thatch, worth 50,000 (INEGI 1995:207), as well as railroad ties, medicinal plants, ornamental plants (a major export), game, wild fruit, charcoal, firewood, local construction materials, and much more. The vast majority of these latter benefits cannot be reckoned; no statistics exist. For example, we cannot even estimate the benefits of the medicinal herbs; how many lives do they save? We do not even know how effective they are. Similarly, there is no way to estimate the quantity of firewood gathered, let alone its benefits to the economy.

Quintana Roo has a great deal of degraded forest left in the south and north, but almost all the commercial wood is in the center and southwest. Yucatán state has cut itself out of commercial production; there is nothing left but minor local cutting—still, alas, enough to prevent regeneration of the surviving forest there. Commercial production of nonprecious woods is insignificant on a peninsula-wide scale, but locally important for construction. The switch to masonry houses has taken much pressure off the local wood resource base.

The problem of logging is, as everyone knows, increasingly intractable throughout the world. Demands for wood are steadily increasing; forests are shrinking. In Mexico, as in most of the world, pasture and farmland expansion is a far more serious threat to the forest than is logging. However, the two processes work together. When the valuable woods in a tract of forest are logged out, the remaining forest is not currently worth much and is apt to be converted into low-grade, low-value pasture and cropland. This destroys its potential as an economic resource; reversing the process, to make it a productive forest again through reforestation (or even natural succession), would require decades. Cutting such high-graded forests destroys biodiversity—including the game and forest plants that are essential to the Maya economy.

Quintana Roo's exploding population presses hard on the forest base. There is no primary forest in the state, and very little remains that is old growth. It is very sobering to carry out research in the magnificent relict forests of Nueva Loria and Naranjal Poniente, or the even more magnificent forests of Tres Garantias, and to realize that most of Quintana Roo was once covered with such. This is not just a matter of tree size. The old forests are more diverse in species. They are multilayered, with at least three tiers rather than the one or two tiers of even quite old second growth. Birds serve as indicators: the Nueva Loria forest is a northern outpost for high-forest species like the eye-ringed flatbill flycatcher, White-breasted Woodwren, and Lesser Hermit Hummingbird. These birds occur only where the forest is multilayered, and are now lacking in most of central Quintana Roo.

At present, only about 50 percent of Quintana Roo is forested and potentially useful for forest products; there are more than a hundred tree species (Kiernan and Freese 1997:94, quoting Plan Forestal data).

Quintana Roo has successfully avoided the unspeakable devastation visited on tropical forests in most of the world. Tabasco has converted almost all of its forest into poor-quality cattle grazing, to its enormous loss. Nearby Belize has been invaded by Malaysian timber companies; the Maya there are protesting this (*Sierra*, March-April 1997, pp. 26–28). These are the same companies that have turned Malaysia from 90 percent forested to almost 90 percent deforested. These companies simply destroy the forest totally, often for almost no financial return—they sell the finest-quality hardwoods, trees worth thousands of dollars per tree on the American woodworking market, for concrete-construction molds, pulpwood, disposable chopsticks, and other low-value uses. They are set up to clearcut and process masses of trees, not to cut selectively; it is strictly a volume operation. They operate with government support or collusion (the corruption in Malaysia, where I carried out research before shifting to Mexico, was not even hidden) rather than in a free-market environment. Typically, the potential free-market value of the wood cut is orders of magnitude greater than the value realized.

All the usual problems of conserving forest are faced by Quintana Roo: expanding agriculture and stock raising, urbanization, overcutting, poaching, forest fire. Overcutting is caused by several factors, sometimes operating together: lack of controls, overly optimistic forecasts, and illegal cutting protected by corrupt enforcement agents (see Ascher 1999 and Williams 2003 for the best world studies of why deforestation occurs). Illegal logging is called *pachocha* (from the Spanish for "stubborn"), and those who do it are called *pachocheros*.

Perhaps the biggest problem, however, is one not often noted in the literature: the problem of underutilized trees. Until recently, there was no market for most species of Quintana Roo trees, and most of the market sales were of only two species, mahogany and cedro. It is highly doubtful if the forest is worth conserving for these alone. Mahogany is uncommon and slow growing. A recent study from Bolivia has powerfully argued this point and raised real concerns about the value of sustainable tropical logging when mahogany is the only economic target (Rice, Gullison, and Reid 1997).

In Quintana Roo, other hard, durable woods were—until recently— used solely for railroad ties, which became a major export. Indeed, the small, wiry, rock-hard trees that develop on Quintana Roo's stony soils are often perfectly adapted for ties, and many are too warped and poorly grained for anything else (except fuel). Even so, this and other types of deforestation are no longer justifiable. This is particularly true for those trees that are good for other purposes. Nothing is more obviously insane, economically speaking, than to go to the trouble of cutting a tree and then burning it or turning it into railroad ties and other cheap uses, when the same tree could be sold as cabinet wood for 3,000 pesos or more per cubic meter. A tie uses almost a cubic meter of wood and brings about 11 pesos. Lovers of wood feel real personal grief at seeing this. To them, it is like watching a Rembrandt get torn out of its frame and used to clean the street.

Most of the hardwood species are superb woods for cabinet work, including furniture, veneer, and even art work. I have seen objects made of *chakte'k'ok'* for sale, for up to hundreds of dollars for a tiny bowl, in galleries in southern California's interior-decorator belt. Some woods, such as *granadillo* (*Platymiscium yucatanum*) and *ciricote* (*Cordia dodecandra*), now are worth more than mahogany and cedro. Others, such as *ya'ax 'eek'* (*Pithecellobium keyense*) and the aforementioned chakte'k'ok' (*Caesalpinia violacea*), are closing in fast. At present, about a dozen species have real value. With any kind of attention to marketing, this number could rise to fifty or more. Even the lowly chechem, which was initially logged partly to get rid of it, has found a niche in high-quality furniture making. Its wood is versatile, close-grained, and capable of taking a high polish.

Conservatism and isolation served until very recently to prevent the development of any sort of woodworking industry in or near Quintana Roo. Mexico's quality woodworking and furniture industry is concentrated in Guadalajara—far away over difficult roads. Nowhere in Quintana Roo was there the technology to season the woods and saw them properly. Proper drying kilns did not exist, nor did modern thin-bladed saws that spare the wood; the old wide-bladed saws reduced an appalling percentage of the log to sawdust. Adapting modern technology to the tropical woods has been necessary. Some Canadian mobile sawmills, developed for Canadian softwoods, were sent to the area in a well-meant aid project. They literally tore themselves apart trying to cut Quintana Roo hardwoods.

Until the last couple of decades, only mahogany and cedro were worth cutting for the quality market; they need less seasoning, commanded a high price, and could be shipped to Guadalajara economically or used locally for rougher but still serviceable furniture. Mexico has sensibly banned the export of these woods, and has banned the export of round logs of mahogany from Quintana Roo, to keep mill employment in the state. Lesser woods were not worth the effort. The Guadalajara furniture makers are often forced to import far inferior wood from the United States. (The foregoing all comes from personal research, including interviews with loggers, woodworkers, and timber merchants in Mexico and the United States; but see also Kiernan and Freese 1997:104, Primack et al. 1998.)

Most trees too soft for ties and too hard for papermaking are currently of no use except for local construction. The Maya find ways to use anything and everything, except some of the rare, spindly trees of the forest understorey. Woodsmen know the special traits of each one: the best vine for tying house timbers, the most termite-resistant wood for half-buried posts, the one plant that does not split when its crotches are used for load-bearing forks in saddles or construction.

By 1996, both government and private efforts had begun. Private workshops in Cancun and Chetumal were using local woods to produce truly superb furniture and doors—world-class woodworking. However, by 2002, this process had stalled, and local hardwoods are still unappreciated.

However, a small export market has developed in other tropical hardwoods, such as granadillo, that are comparable in quality with the precious woods but are not covered by the export ban. The wood often goes to East Asia, where its beauty is more fully appreciated than it is in North America. North American woodworkers would, however, gladly pay top prices for properly seasoned Quintana Roo hardwoods. These woods are as beautiful, durable, fine-grained, and versatile as any in the world. Many have unique color and pattern. Only two links—modern seasoning and sawing facilities (with quality control), and marketing infrastructure—separate those ties from North American woodworkers who would pay thousands of dollars per tree.

A larger—if more humble and utilitarian—market has developed for the woods with somewhat less potential (see Kiernan and Freese 1997:112–13). *Pukte'*, ja'abin, and other woods that are unlovely but rock hard, make superior pallets and framing boards. Tsalaam makes superior flooring, like hard and exceptionally fine-grained oak, and is widely used for this purpose; many planks from Chunhuhub become floors in Mexico City and other urban areas. (They were selling at 7 pesos per board foot in 1997, and more by 2001, by which time they had become the mainstay of the town lumber mill.) Fast-growing, pithy softwoods can be used for paper or for minor items from dowel rods to throwaway cement facing. A brief boom in logging for pulpwood took place in 1996–97, but when the more easily accessible softer-wooded trees were cut, further production became uneconomic.

Last of all, there are all the nontimber products of the forest to consider. The Maya need thatch from palmettos (xa'an; *Sabal* spp.). A xa'an roof must be replaced every twenty years at the outside, and hurricanes can force the replacement of almost all the thatch in a town. One leaf costs a peso or more. *Juulok' xa'an*, particularly broadleafed individuals, are especially regarded. Palmettos are saved when milpas are cut, but forest-grown leaves seem preferred. Forests also produce bajareque poles, medicinal herbs, and many other vitally important goods. The people of the area depend on the forest for construction materials. Most houses are still pole-and-thatch and require considerable raw material. Builders must know what wood is good for what purpose. The high forest is particularly valued; poles from high forest are tougher than those from second growth. However, most people must make do now with the latter. A house 7 x 4 m costs 3,000 pesos (1996) for material, and 2,000 to build, if one hires builders.

For the main posts, the most valued wood is ja'abin, because it is hard and durable and extremely resistant to termites and dry rot. Sapote is also excellent wood for posts and beams but is rarely available, because sapote trees are not normally cut. They are more valuable alive. Used for posts also, but especially for beams (as the name of the first tells us), is *chakte'viga*, also called *k'anchakte'* (species of *Caesalpinia*).

Traditionally, the house was tied together with tough vines, not nailed. These vines were of the general sorts called *bilunkok* and *anik'aab*. These names are cover terms for a number of lianas in the family Bignoniaceae.

Furniture for use and sale is made from the standard valuable woods—mahogany, cedro, granadillo, and others.

It is clear that the mixed forest is more valuable than anything that can replace it—so long as the mixed forest is fully valued and utilized. It is the perfect crop. Tree farms cannot replace it; they cost a great deal to operate, and plantations have serious pest problems because of monocropping. Maize and cattle have their place but will never thrive on the stony flatlands of central Quintana Roo or the high, steep, rocky ridges of the southwest. Every agricultural experiment in these areas has ended in disaster. Even subsistence milpa does poorly (it does better on the dry hills and in the alluvial valleys, which are not very good for forestry).

The problem is one of retooling the extractive economy to use all the common trees, not just two or three of them, and also to value the harvest of medicinal herbs, the game, and the other resources of the forest.

The two real mainstays of forestry in Quintana Roo are mahogany and cedro, the "precious woods."

Mahogany (*Swietenia macrophylla*) is well known throughout the world as one of the most beautiful and resistant of all woods. The tree is a beautiful, wide-buttressed, straight-boled tree that grows fast to a great height. In Chiapas, at least, it tops 70 m, but rarely much more than half that in Quintana Roo, where conditions are worse and remote, uncut forests do not exist. (Soon they will not exist in Chiapas, either, but that is another story.) Even so, I have seen trees almost 2 m thick and a good 40 m high. The major problem of mahogany as a timber tree in central Quintana Roo is heartrot, to which it is very susceptible. I have seen a magnificent tree 2 m thick that had to be abandoned; on felling, it proved to be a mere shell around a rotten core. It should have been left as a seed tree, but there was no efficient way to determine its condition before cutting it.

Under the worst conditions Quintana Roo can offer, mahogany grows only 0.33 cm per year, but, in Chunhuhub, even on dry hills it does better than that. Chunhuhub woodsmen expect it to grow up to half a meter a year. I have seen it grow a meter a year on really good sites. Mahogany cannot be cut until it reaches 55 cm dbh, as opposed to 35 cm for ordinary hardwoods. (Dbh, "diameter at breast height," is normalized at 4.5 feet above ground, and is the usual way to measure forest trees. This can be a bit deceptive in an old tree because of the buttresses). This size is reached in about 60 years or less on Chunhuhub sites, but not until 100 to 120 years of growth on some of the windswept, rocky lands of east-central Quintana Roo (Snook and Barrera de Jorgensen 1994).

Mahogany is a light-gap tree: it loves gaps in the forest (Snook and Barrera de Jorgensen 1994). Snook maintains that it prefers large gaps with disturbed mineral soil, but this is not the experience of woodspeople in the Chunhuhub area. In Chunhuhub and Naranjal Poniente, mahogany trees require light gaps to grow, but they prefer small ones with reasonably good soil, not the big ones produced by modern cultivation or even the larger traditional milpas. They suffer from accidents such as wildfires (an escaped milpa fire burned forty young trees on the main road in spring of 1996). They do extremely well along roads and other small, partly shady, humid openings. Unfortunately, roads are bad places to plant mahogany—any widening or clearing project takes them out unless carefully done. Also, the trees are almost invariably stolen as soon as they get anywhere near marketable size. Cutting of undersized mahogany is universal in isolated parts of the state. Road crews carefully clear around mahoganies planted by reforestation projects along the highways, but this merely calls attention to them. Some individuals plant mahoganies in isolated milpas before abandoning them, but most milpas are recut on a rather short cycle, making reforestation impractical.

Mahogany loves nothing more than the traditional conditions of the Quintana Roo forest: a dense, thick, fairly wet forest where blowdowns, small burns, and other small openings are always appearing. Its light, wind-distributed seeds snow down from old established stands, and the seedlings quickly fill these small gaps. The established trees live for decades or centuries, eventually coming to dominate the forest. Unfortunately, the whole strategy of the tree is premised on a world of many small light gaps and much undisturbed time between local windfalls. With modern large clearings and selective logging, most of the habitat is poor and most of the seed trees are gone. So mahogany cannot reproduce successfully in most of Quintana Roo. Its future depends on reforestation and protection of the stands, and on people's ability to plan for the long term.

The model for this in the immediate area of Chunhuhub is Naranjal Poniente, a small village with a huge land base, some 30 km south of Chunhuhub. This town is blessed with a superb tract of old-growth forest, on flat rocky ground but with well-drained fertile soil. Here, mahoganies rise high above the lesser trees and dominate the skyline. It is probably a good sample of what Quintana Roo was like before the selective cutting of the last century or so; mahogany vies with sapote and ramon for dominance in these forests. The ejidatarios have worked closely with the Plan Forestal, a government plan for cooperative management of forests (see below, and Anderson 2003). Logging is selective and conservative, under careful controls. The ejido plants at least three for every one cut. I walked all through the forest with my Maya friends, and we found no evidence of illegal cutting.

The only seedlings in the dense forest are sapote, ramon, and smaller trees, however. The mahogany trees must commemorate natural or agricultural burning. Ejidatarios are opening up the forest and planting mahogany to keep it growing there.

The government has controlled logging on many of the ejidos that still have rich mahogany supplies. To some extent, this has merely displaced heavy logging across the borders into Belize and Guatemala, which supply a good deal of this wood to Quintana Roo mills (Kiernan and Freese 1997:102).

The other precious wood is cedro (*Cedrela odorata*, sometimes called "Spanish cedar"; since it is neither Spanish nor a cedar, "cedro" is a better name). Its similarity to cedar is solely in the wood: the tree has a light but strong wood, fibrous and tough but lovely and easy to work. It was well known to an earlier generation of Americans as the raw material of cigar boxes. (I remember from my early years how many uses there were for cigar boxes after the cigars were gone; since the wood was so tough, the boxes lasted indefinitely and were recycled in countless ways.) Though less lovely and durable than mahogany, it is pest resistant and versatile and brings as much on the market.

Cedro was never very common in Chunhuhub and has now been eliminated as a timber tree, but it abounds as a shade tree in the town. At least for local uses, such as furniture, it should be available in future.

The Maya word for cedro is rather thought-provoking: *k'uj che'*, "divine tree." The specifics of its divinity are lost in the mists of time.

Cedro, though related to mahogany, is a tougher tree. It too is a light-gap species, but it prefers bigger gaps with less fertile soil; its idea of paradise is a village street. It is, in fact, one of the commonest weeds in Chunhuhub. Its seedlings quickly fill any abandoned garden or undisturbed roadside. It has serious problems in the small, humid, vegetation-choked gaps that mahogany favors; it much prefers a sun-baked expanse of bare soil, where it can grow without much competition.

Cedro, therefore, it is a reforester's dream. Weed tree that it is, it grows up to 2 meters a year, and it survives conditions that have to be seen to be believed. Unfortunately, it shares with mahogany a susceptibility to boring insects (especially *Hypsipila grandela*) that often kill the terminal bud. This does not hurt the tree very much; it just keeps growing from lateral branches. However, it spoils the beautiful clear boles for which forest-grown cedro is famous. The insect is not a great disperser and fails to find trees out in the forest; but in plantations it is devastating. Appropriate control measures seem difficult to devise.

Even so, reforestation projects are under way. The town nursery has grown and planted out a great deal of it. At the García family ranch, Rancho El Corozo, for instance, there are several rows of young cedros. They were planted on a dry but fertile terrace and intercropped at first with chiles, which do well enough there to provide a good cash source while the cedros grow. As of 1996, some trees had been in the ground four years and were about 6 meters tall. Some showed signs of forking, but the boring insect was not a serious problem—yet. Cedro planting is popular for smaller-scale farmers too, including subsistence cultivators in Presidente Juarez. Sometimes the government pays them to plant.[1]

The infertile soil of the flatwoods is now generally left under forest, carefully managed to produce a variety of valuable woods on a fifty-year rotation. It is possible to cut milpa, crop the land for two years, and return the forest (with reforestation). However, current regimens of selective logging are preferable. They ensure soil protection, and they allow small-scale reforestation in beneficially small gaps. Most valuable to the local residents, they create a steady, reliable, predictable income stream. By contrast, a plantation-type system on a fifty-year cycle, or any other cycle, would produce income only once in many years—and then only with luck. All too probable would be a situation in which the trees grew for forty or forty-five or forty-nine years, and then were destroyed by fire or hurricane just before they reached paying size.

Until recently, isolation and low population protected the forest. Today, this is no longer the case. Population pressure is forcing the full utilization of the Quintana Roo landscape.

Chunhuhub, which in 1991 was a model of planning for forest conservation, had eliminated its (admittedly very small) reserves of valuable woods by 1996. Weak leadership allowed overoptimistic projections and led to failure in enforcing the rules. I know the leaders in question, and I believe them to be honest men; the problem was that no one would take an aggressive initiative in the rather dangerous activity of stopping poachers. The result was ruin: the ejido was one of the richest in Mexico in 1991, due to sale of precious woods, but was virtually without income in 1996. This, of course, foreclosed the possibility of fixing the situation; reinvigorated planning, enforcement, and reforestation would have taken a great deal of cash.

Meanwhile, the town government in 1997 came under the control of an individual known for his close ties with illegal loggers of precious woods. Conservation and planning came to a definitive end. Failure of solidarity in the community had prevailed, in spite of both governmental goodwill and local desire. Lest readers be tempted to blame the local capitalist elite, it should be reported here that one of Chunhuhub's major figures in commercial agriculture told me with open disgust that the town's political leaders were "too lazy" to plan or to work on the problem; the lumber mill owners, too, were hardly in sympathy with someone who was in a fair way to ruin them.

By that time, Quintana Roo had very prudently stopped issuing logging permits to ejidos that did not have plans for sustainable logging.

(Would that other polities, the United States notably included, had similar values and common sense.) Chunhuhub was thus barred, briefly, till government and planning changed, from cutting anything at all.

First World environmentalists, secure in their relatively well regulated homelands, do not understand the problems faced by tropical forest towns. First, the citizens have to be convinced that the forest is valuable. The Quintana Roo Maya need no convincing on that score, so that half of the battle is won already. However, the citizens then have to get together and work as a body. They have to plan realistically. They have to come up with a practical enforcement structure. And they have to work well enough with higher levels of government to be successful at "comanagement."

Chunhuhub came agonizingly close to success. But conflicts between Chunhuhub's leadership and higher levels (including some officials who were perhaps not models of probity) led to leadership changes, as we have seen.

Moreover, Chunhuhub's leadership elected not to work with the Plan Piloto Forestal (see below). This proved to be a bad mistake. At first it seemed reasonable. The Plan charged a fee for services, and it did not seem obviously a better or more reliable agency than countless other agencies that had been forced on the Maya in the past. The usual story of agencies, plans, programs, and projects in Mexico (as elsewhere) is simple: the project starts with much fanfare. Local citizens contribute money and/or time. Officials enrich themselves and leave, abandoning the project (cf. Hostetler 1996).

Hundreds of thousands of hectares of Quintana Roo have been cleared, at one time or another, for projects that were at best impractical and at worst insane. I was shown one large tract of land in a nearby ejido that was cleared "for agriculture" in the 1970s, and duly written up in the statistics of that period as "agricultural land opened." A rich, productive forest had been cut. Nothing else had been done. The area was waterless. Irrigation service was promised but never provided. The soil soon eroded away, and when I saw the place (almost twenty years later), it was a neatly squared desert.

A forestry experiment was the Maderas Industriales de Quintana Roo (MIQRO), a gigantic project that began in 1954. Intended as a state-controlled model logging project, MIQRO focused on only two species, mahogany and cedro, and high-graded the forest mercilessly. It cut out the two species from its vast concessions, and no one thought to develop uses of other woods. The high-graded forests were often cleared for agriculture. Other benefits to local people were few. (My interview data and Kiernan and Freese's review [1997:96–97] are in agreement on these points—always a cheering result in research.) MIQRO's giant sawmill became inactive.

But the Plan Piloto Forestal was different. Its story has been told in a superb (but, alas, unpublished) thesis by Julia Murphy (1990) and briefer but excellent and fortunately available reports by Michael Kiernan and Curtis Freese (1997) and Henning Flachsenberg and Hugo Galletti (1998; Galletti 1998). Since they wrote, the Plan has become accepted, indeed is a fixture of local management, and is thus no longer "piloto"—it is now the Plan Forestal.

The plan was developed by federal and state officials and biologists, with German aid and support, and now Germany preferentially buys tropical wood from such ejidos as have developed plans for sustainable harvesting, such as Nohbec. (Germany has an enlightened national policy of preferential buying of tropical woods from communities that have developed such plans.)

The Plan Forestal involves ten ejidos and several private property owners. The latter, who are owners of what would be called "wood ranches" in the northwestern United States, form a separate organization. The basic activities of the Plan are to supply technical advice (for a small fee) to the local people, and to help them work out management plans for their forests. Ideally—and, to a great extent, in practice—the Plan officials provide the biological and industrial knowledge, while the local people provide the working knowledge of the forest and its trees and make the final decisions.

Plan Forestal documents explain how to plan twenty-five-year rotation, work with ejidos, cruise for trees and mark them, set up an ejido woods boos system, and even how to cut lianas so a falling tree falls clear instead of getting caught in a cat's-cradle of tough vines. All this is in simple language and laid out in cheap mimeographed papers. The whole is a model of what can be done.

Another example of what can be done comes from the Plan's work with luthiers, who are now using Quintana Roo chechen for guitar fingerboards, and mahogany for the bodies and necks (*The Canopy*, newsletter of the Rainforest Alliance, March-April 1998, p. 1; Jenny Erickson, personal communication). This not only provides a market for the chechen (not the most popular tree in the Yucatán), but provides a use for branches, trimmed pieces, and other small parcels of wood that are now generally discarded.

Kiernan and Freese fear a future in which the cut will have to be reduced to so low a point that the ejidatarios will not want to keep the forest. I am less worried about this. The Maya want their forest, and not just for financial reasons. I am more concerned with the prospect that a later and less idealistic generation of Plan Forestal officials might try to dictate to the Maya, instead of working with them. Given the opinions of the Quintana Roo Maya about such attitudes, this would bring about the downfall of the Plan Forestal. In the meantime, the Plan Forestal is an eminently successful example of comanagement. It has, so far, avoided the great problem of comanagement regimes: the insidious tendency of top-down control to reappear. This may happen either because of international political-economic forces, or because mistakes at any level are "corrected" by adding another top-down rule.

Richard Caulfield has recently written a thoughtful book about Greenland whaling that makes these points (Caulfield 1997). Whales and mahogany trees may seem different, but they share some important characteristics: they are rare, they are valuable, and they are easily overharvested. Greenland Native peoples find themselves riding a razor edge. If they overharvest, they bring the wrath of central administrations and of worldwide opinion down on themselves, and they find themselves subjected to more and more restrictions. But heavy-handed restrictions lead to more poaching.

The only solution for the Maya is to expand uses of the forest, and to reassert their role as planners and decision makers, while listening to the advice of biologists. In the future, it will probably be necessary to develop some sort of local enforcement system that will be able to deal aggressively with illegal logging.

The Plan Forestal, and even its ill-fated predecessors, had the good sense to insist on selective cutting. Clearcuts in Quintana Roo are an unqualified disaster. Small clearings, such as milpas, grow back quickly—partly because the Maya coppice trees in the larger milpas rather than eliminating them. Large, totally cleared areas, however, are exposed to the full force of the sun and rain, and are isolated from seed sources of the more valuable trees. Moreover, they grow up to tall dry weeds and grasses, which invariably catch fire and destroy every new tree. They thus become worthless and barren for years, and sometimes never recover.

Kiernan and Freese (1997:124) point out that the "central-government, top-down approach . . . has clearly failed in Mexico." Control is impossible on the ground, and the officials in charge of enforcing it are easily corrupted by payments from unscrupulous mill owners. Kiernan and Freese go on to note that the bottom-up approach rests on four points: secure tenure for the local people, direct negotiating power for those same local people, equitable distribution of benefits, and development of (or use of) local expertise for decision making. One must add that the local communities must have recourse in case of overbearing officials, and that the local officials must be carefully chosen.

The bottom-up approach of the Plan Forestal is successful for a number of reasons. First and most important, the people on the ground are inevitably those who make the real decisions about resources; in a top-down situation, the top tries to "control" them, and they respond by turning to pachocha, but in a bottom-up situation, they are under maximum pressure to control their own behavior. Second, a project like the Plan Forestal attracts a superior breed of administrator. It is idealistic enough to interest the most responsible and gifted people, and poor enough financially to discourage the corrupt. The North American belief that all Mexican officials are corrupt is absurd—a mere stereotype. The problem is that those who are corrupt are maximally able to take advantage of opportunity, largely because of structural reasons, especially the above-mentioned problem with recourse mechanisms.

The Plan Forestal involves only a small percentage of the land in Quintana Roo, but its effects extend well beyond that land—a point that needs to be made in evaluating its success. Its principles and practices have not been lost on the Maya, quick as they are to observe and evaluate every new development in the forest. The Plan Forestal has made every forested ejido in central and southern Quintana Roo think more seriously about conservation and long-term management. The Plan Forestal ejidos are leaders, and everyone is watching their progress.

One poster-child ejido for the Plan Forestal is Tres Garantias, lying in the far south of Quintana Roo. Tres Garantias is a truly incredible place, for a number of reasons. It is one of the last unspoiled pieces of Quintana Roo. Its enormous extent—fully 43,040 ha—is inhabited by only a few hundred people. There are 105 ejidatarios, which means a population of about five hundred to six hundred. (Only heads of families are ejidatarios.) Its people are half Quintana Roo Maya and half immigrants from elsewhere, mostly nearby Campeche and Tabasco; yet the people all get along well, cooperate, and understand each other. They share a common Maya heritage at some remove.

Tres Garantias lies in the southwest corner of Quintana Roo. This is a land of long limestone ridges that run north-south. From a high post, the land looks like a vast sea, its great green rollers seemingly blown by the east wind. The forest is deep and tall, far taller than anything in the Chunhuhub area. The canopy ranges from 30 to 50 m. Mahogany, wild fig, ramon, and sapote are the real forest giants. As usual, the latter two dominate in very old growth forests where trees have had to come up under shade. A striking feature of the forest is the abundance of huge old allspice trees, some of them fully a meter thick. Old allspice trees have bark like a sycamore's, beautifully mottled with tan and silver. The most magnificent forests seem to be on north slopes. Small ponds lie in the valleys between these tall, steep ridges.

Tres Garantias sadly destroyed much of its forest in ill-advised and disastrous agricultural schemes. Thousands of hectares were cleared for cattle, but the land was unsuitable for pasture. Anyone with a knowledge of tropical ranching would have known this, but these forest Maya and mestizos did not; they listened to the "experts." Crops did poorly because of dry, phosphorous-poor soils; much of the land is low-lying, and acid in the soil binds what little phosphorous exists. Peanuts and beans needed irrigation, and the pumps wore out; chiles succumbed to whitefly, with pesticides too expensive to be a solution; bracken fern took over much of the milpa land. Gramoxone was tried as an herbicide, and it seemed to favor the bracken and the altanisa (a common weedy composite) over the crop plants (so the villagers told me).

Unlike most of the world, Tres Garantias learned from its mistakes. Half the ejido is now reserve. To be exact, 20,000 ha have been reserved, in cooperation with the Plan Forestal. Of this, 15,000 ha are used for sustainable forestry and for subsistence hunting and gathering, 5,000 ha form a wildlife reserve; hunting is forbidden. A 400-ha core of the latter is deemed so valuable and vulnerable that it cannot be entered at all by outsiders.

In addition to carefully controlling the cutting of mahogany and rarer hardwoods, and carefully controlling the hunting of deer, peccary, and other animals (Anderson and Medina Tzuc 2004), the ejido worked with Plan Forestal personnel to develop ecotourism. There is a tiny lodge in the wilderness, at the end of a long jeep trail; here tourists can stay in small groups and see crocodiles, monkeys, and peccaries. Tapirs and jaguars are present, but are harder to spot. Plan Forestal is also using some of the fees it charges to develop handicrafts in the area.

Traditional beliefs help safeguard the forest. Federico Gutierrez, a Maya from Yucatán and now an ejidatario of Tres Garantias, explained to me that the Yumilk'aax—the Lord of the Forest; unlike mana Maya, he thought there was only one, or at least only one leading one—punishes overhunting and overtapping of chicle. The Yumilk'aax, according to him, is *Dueño de la Selva* or *Juan de Monte* in Spanish. Don Federico has a strong traditional conservation ethic, newly strengthened by the biological science he has absorbed from A. Ehnis Duhne and other tireless workers associated with the Plan Forestal projects. For him and the ejido, the experts tilted the balance. It is highly doubtful whether the experts could have succeeded without the traditional conservation ethic. It is certain that the conservation ethic could not have succeeded without support from the government; there are those ruined pastures, and on the way back to Chetumal one passes through vast ruined landscapes in which local conservation ethics could not stop lunatic development schemes.

The Plan Forestal is about as ideal a solution as one can reasonably expect. However, to succeed, it must be combined with far more aggressive development of processing and marketing, especially in regard to the "other" trees and forest products.

Elsewhere in Mexico and the Caribbean, many of these are extensively used. Wild fig and chakaj, for instance, are major wood sources for handicrafts (see the excellent and important article on this by Peters et al., 2003—a model of how to study and plan for Mexican wood use). The future of Quintana Roo's forest—if it is to have a future—lies not only with Plan Forestal, but also with those firms in Cancun and Chetumal that produce top-quality furnishings. It lies with the medicinal herb garden near Felipe Carrillo Puerto—alas, destroyed prematurely by Hurricane Roxane. It lies with the peccary farm in San Francisco Ake. It lies with the new plantations planned in various parts of Quintana Roo for softer-wooded trees for industrial uses. There are very few plantations in tropical Mexico—a little pine (*Pinus caribaea*, native to Belize and points south), eucalyptus, cedro, and the like. All these projects must support each other. Comprehensive planning and comanagement must be supplemented by really serious attention to marketing Quintana Roo's high-quality woods and the products that can be made from them.

In September of 2001, Chunhuhub overcut its allotment of precious wood.

Before this, conservation and reforesting in the past had paid off. After the exhaustion of precious wood, in the early 1990s, young mahogany had grown up. Some had finally reached the magic size of 55 cm diameter (breast height), big enough to be cut.

Harvesting duly began, and there was an overharvest of a tree or two. Some thirty people petitioned SEMARNAP (Secretaría del Medio Ambiente, Recursos Naturales y Pesca)—Mexico's department of the environment—with a complaint. PROFEPA (Procuraduría Federal de Protección al Ambiente), the enforcement arm of Mexico's environmental protection system, cracked down, investigated, and found an overcut of 35 m³. (The permit had allowed 1,200 m³, so this was a relatively small overrun.) Punishment was threatened. There was much rancor toward the whistle-blowers. Two noisy and fiery assemblies of the ejidatarios followed. The ejido treasurer, an educated and competent young man, managed these, providing the voice of reason. The ejido pled with PROFEPA for forgiveness on promise of reforestation and good behavior.

This story reveals a whole string of enormously important recent developments. In the late twentieth century, a totally different attitude toward Quintana Roo's resources surfaced. People began to realize that Quintana Roo's natural treasures were desperately threatened and were valuable enough to be worth saving. The previous attitude—typical of the entire twentieth century up to 1990—had been that there was plenty of game and timber, and in any case Quintana Roo should be deforested and turned into pasture and agribusiness land as fast as possible.

The change of heart in the last five years has been dramatic. The idea of limits has gotten through. Even older individuals who cannot quite adjust to the new reality are aware that the government has reversed itself on these issues. The young are much more aware of the reality behind the change. They have grown up in a world of wider and longer vision—and of more obvious limits. The exhaustion of Chunhuhub's precious woods in the early 1990s was a wake-up call.

Knowing there are limits is meaningless unless one also knows that there is some recourse—that the destruction of nature is not inevitable. Until recently, the general attitude was that modernization must necessarily destroy the local and natural, and substitute for it the introduced and artificial. The idea that local natural resources could actually be saved, and managed for sustainable cropping, simply did not exist in most of south Mexico (and many other tropical lands). Today, people are aware that there is recourse: the resources can be saved, and there are mechanisms in place for doing that.

The Mexican government, especially at the national level, has shored up environmental protection considerably. Until very recently, enforcement of environmental protection laws was not taken seriously in Quintana Roo. Enforcement officials were few. My interviews showed that they were often unaware of the laws. Also, they were sometimes ready to accept even quite small *mordidas* ("bites"—bribes) for overlooking major infractions.

Today, patrolling and enforcement are notably better than they were ten years ago. Patrols from PROFEPA are visible, professional, and frequent. Big operators in Cancun may still get away with dumping raw sewage into the lagoons and building hotels in theoretically protected and highly sensitive areas, but the Maya do not have the political power to do such things. Indeed, even Cancun is beginning to feel the pressure, sadly late in the game. Some areas, such as hunting, remain uncontrolled, but timber is important enough to the Quintana Roo economy to attract some serious protective attention.

Especially important has been the success of the Plan Forestal. It convinced many people that conservation was not only a good idea, but it was actually possible. Previously, much of the problem had been one of cynicism. Earlier conservation plans had never been enforced. The view arose that one simply could not save trees. This view persists among most people today about game. However, even in regard to game animals, the clear success of sustainable management in other areas has convinced people that conservation is not just another fly-by-night plan. Reports from elsewhere in the peninsula (e.g., Alcocer Puerto 2001; Almanza Alcalde 2000) make it clear that SEMARNAP, PROFEPA, and the Plan Forestal are beginning to drive home a message that the traditionally conservationist Maya are now ready to hear.

A ban on cutting old-growth forest has been invoked, although it is not universally enforced. Theoretically, milpas must be cut in younger woodland. In 2001, I saw more than one milpa cut in old forest, but at least the rate of destruction has visibly slowed.

Tie cutting, so serious a problem in the early 1990s, has dropped to a low level. Mexico has finally realized that the trees are worth more as standing forest, let alone as actual commercial lumber. The railroads have switched to concrete ties.

Minor hardwoods are now being used for furniture, flooring, and other specialized uses. Many Chunhuhub homes sport beautiful carved doors, chairs, and tables made by local craftsmen from local woods. Market failure both deprives the First World of such wood and such craftsmanship, and deprives the local people of the spectacular returns they could get for their efforts.

Reforestation has become almost a craze. Huge and well-run government nurseries are turning out vast numbers of seedlings, mostly cedro and mahogany. People are paid a small sum—enough to be attractive to the Maya—to plant these, and many do it, especially in the usual tree-planting season (late fall). Young trees are growing up everywhere. Cedro seems to be more a plantation tree; mahogany is everywhere—along roads, in windbreaks, in forest openings. Good advice is available—for example, to tell the Maya not to pull up saplings by the roots to transplant them, a bad habit though not necessarily fatal in the wet, rich soils.

Finally, and most important, there was a forest reserve of about 1,200 ha as of 1996; by 2001 the town had been persuaded to designate its entire 10,000 ha of surviving old-growth forest (a huge area of flatwoods) as reserve.

With all this, the future of forestry in Quintana Roo can be viewed with cautious optimism. If the government continues to advance in its comanagement programs, working with Maya ejidos and individual land-owners, the future will be bright. Weak will or, far worse, a change toward a top-down or large-scale exploitation policy will lead to destruction of the Quintana Roo forests. Meanwhile, the tourism economy is beginning to move from the shorefront into the forest areas of the state, and serious damage could result in the future. Eternal vigilance and genuine commitment to sustainable forestry are the price of economic liberty.

In the meantime, the Maya care about and care for their forest. On one of our research trips, Don Felix cleared a little firebreak around a *caobita* (little mahogany tree) in the forest. He had no reason to do so; he would never see it again, let alone profit from it. His use of the affectionate diminutive explained his motive well enough.

CHAPTER FIVE

The Lords of the Forest
Ideology and Biotic Resource Management

CARLOS LENKERSDORF has argued that the Maya language encodes a distinctive worldview (*cosmovisión*, which could mean "cosmic vision," but here means "worldview" as conceived by Michael Kearney 1984). In a recent book, *Los hombres verdaderos: Voces y testimonios tojolabales* (1996), he has elaborated this view as he understands it. He is writing about the Tojolabal of Chiapas, but his conclusions should apply to the Yucatec as well, since the languages and worldviews are closely related. The Tojolabal have, however, a very different history from the Yucatec of Quintana Roo; a small, vulnerable group, they were early reduced to poverty by Spanish land grabs, and were never able to reassert themselves. They have thus remained more traditional in some ways—having been prevented from rising in the world, educationally and economically—while being at the same time less able to resist outside influences (or so it appears from my understanding of Lenkersdorf 1996, and from personal communication by Jan Rus [1996, 1998], who knows the area as do few outsiders).

Lenkersdorf finds that the Maya languages embody an idea he calls "intersubjectivity." Maya languages are ergative. In English-language terms, this means that the object of action is featured in the sentence. "I hit the ball" becomes "The ball, I hit it"—but expressed grammatically with a distinctively transitive verb form and with the subject "I" reduced to little more than a prefix of the verb. In Yucatec, and I gather in Tojolabal, the pronoun form used in this context is essentially a possessive form, so one is saying something like "The ball, my hit it." Or consider the remark that gave Yucatán its name: "matan cub a than" (Restall 1998:122), in modern spelling *ma' a t'aan, ku y u'ubik a t'aan*, "he doesn't understand what you said"—literally, "doesn't your words, he understands your words" (with "he" in a possessive form, parallel to "your"). The object goes first—after the negative particle *ma'*—and is repeated at the end; the verb is in the transitive form.

Naturally, the distinction between transitive, intransitive, and passive verb forms is fundamental in an ergative language in a way it is not in English or other non-ergative languages. Yucatec has a separate declension system for each of those three modes. The verb dominates the sentence in a way not found in subject-object languages like English, with their obtrusive topic subjects.

Lenkersdorf argues that, though the ball is the "object" of the action, it is the main *subject* of the sentence. It is clearly the thing the sentence is about, and presumably the thing the speaker is interested in. Other linguists prefer the term "patient," which underlines the fact that this is *not* either a "subject" or an "object" in the English or Spanish grammatical sense. The actor then becomes the "agent," not being a true "subject." For Lenkersdorf, the sentence has two subjects—the agent and the patient are both, in some sense, subjects. Lenkersdorf sees Maya grammar as embodying an idea that both the agent and the patient, and anything else central to the action in the sentence, are interacting subjects—or interacting principal topics, to translate more freely.

Lenkersdorf sees a whole philosophy expressed in this intersubjectivity. He finds the Tojolabal existing in a world of persons who are equal (or close to it) and mutually responsible. He sees the Tojolabal community as egalitarian, prosocial, and interactive.

This becomes part of cultural ecology when the Tojolabal deal with nature and farming. They see nonhuman species as "subjects" too, not mere objects to use and throw away. They respect trees and animals; they respect the soil and have a reverence for the *madre tierra*, Mother Earth. They personify and personalize the natural world (Lenskersdorf 1996:71–72).

Lenkersdorf contrasts the Tojolabal view with the "Indo-European" view, which he sees as characterized by power relations in a hierarchy of domination—as expressed by subject-verb-object grammar. He sees the Tojolabal as trying to preserve their egalitarian, homogeneous communities in the face of domination and exploitation by Spanish-speaking society.

The dark side of this intersubjectivity is enforced conformity; individ-ualists and nonconformists are *pilpil winik*—self-separated people. They may find themselves ostracized or banished. Lenkersdorf discusses the costs and benefits of this from both Spanish-speaking and the Tojolabal standpoints. The European mind sees violation of human rights; the Tojolabal sees maintenance of the intersubjective community, the proper universe of discourse. (One wonders if the Tojolabal—or, rather more likely, Lenkersdorf—have been reading Jürgen Habermas 1984.) The pilpil winik apparently see things more or less as the Europeans do, not as their more conformist fellow Maya do. Perspective may matter more than language.

There seems little question that Lenkersdorf is a romantic of the old school, idealizing the happy community of farmers, just as the ancient Greeks and Romans did two thousand years before. It is also clear that he is drawing his comparison in the starkest possible terms. Like those Greeks and Romans, he has a satirical intent—he is not only describing the Tojolabal; he is holding his own Hispano-Mexican society up to a mercilessly revealing looking glass! In fact, the Tojolabal are rather less united and harmonious than Lenkersdorf allows (Jan Rus, personal communication 1996, 1998). And certainly the Spanish-speaking society of Mexico is not so bleakly fixated on dominance and exploitation, or Lenkersdorf himself could not have emerged from it and won a prize for his book criticizing it. Also, a linguist would object that there are ergative languages that do not seem to accompany such a worldview and that, more to the point, there are countless "little communities" around the world that are egalitarian, communitarian, and environmentally aware without speaking ergative languages.

All that said, Lenkersdorf makes some important points. The ergative constructions may fit this particular variant of that worldview better than other languages, even if others may speak such languages without losing the worldview. Certainly, there is more than a little truth in his characteriza-tion of Maya thought. The Maya construct an entire worldview based on intersubjectivity.

Lenkersdorf's conclusions for the Tojolabal apply to the Yucatec as well. The Quintana Roo Yucatec have never experienced such ruthless domination as have the Tojolabal. This may explain the more individualistic cast of the Maya of Quintana Roo. By comparison with Lenkersdorf's Tojolabal, the people of Chunhuhub appear to be more individualist, competitive, modernizing, and tolerant of wide differences in economic and personal situations, but they have the same basic worldview: intersubjective and interpersonal, with nonhuman lives taken as part of the wider community. They are intersubjective in Lenkersdorf's sense, but less tightly so; they are a bundle of free strands, not a tightly woven net.

Now that scholars can read Classic Maya, we learn that the Maya of that age were an individualistic, competitive people; most of the inscriptions glorify the deeds of kings and other elite individuals (Schele and Freidel 1990). More recently, the Cruzob Maya of the nineteenth and early twentieth centuries were split into factions, and their society was marked by strong differences in power and authority. They were independent, individualistic, and strong willed (Dumond 1998; Villa Rojas 1945; Sullivan 1989).

This extended treatment of Lenkersdorf's ideas serves to introduce the Maya strategy for managing the environment. Lenkersdorf states a basically accurate position in a strong, clear, uncompromising way. He takes a strongly phenomenological position. The fundamental views of the manager about life, person, and being are the logical starting point for analysis of environmental management.

All too often, phenomenology leads to a sort of individualism that is almost (or even completely) solipsistic.[1] Lenkersdorf (like Habermas) saves himself from this trap by his focus on language and speech communities. This viewpoint emphasizes the fact that people share their views, morals, and ideas through dialogue.

In this case, "polylogue" would be a better word, but I spare the reader such a neologism. The reader should recall, however, that the "dialogue" in question involves many people. Ejido *asambleas* (assemblies) in Chunhuhub typically involve more than a hundred individuals, of whom at least twenty to thirty normally can be expected to voice their views. Far more people are involved in the networks of communication that sustain management ideas and ideologies. Individuals talk to one or a few people at a time, but messages spread with incredible rapidity along the communication channels, and thousands of people are incorporated.

The alternative to looking at worldview would be to start with the actual behavior observed on the ground. A biologist might even ignore the

Maya philosophy of management altogether. Human ecologists sometimes treat verbal report as a bogeyman, serving no purpose except to mislead fools who should better be recording "behavior" (see, e.g., Harris 1968).

However, talking is a type of behavior. Not only that: it is one that is not without influence. Thus, even the most behavior-oriented anthropologists are interested in the dialogic process—in what people say, and in what happens after they talk.

The arguments and debates in Chunhuhub's ejido assemblies are translated into rules and into action. People get land and use it according to assembly decisions. Rules for forest, game, and milpa management are made in these assemblies, and then enforced or unenforced according to how much public sentiment is voiced. This distinction is an important one. When public health authorities directed that pigs be kept out of the streets for reasons of sanitation, the citizens of Chunhuhub solemnly assented. No diminution in street pig population was noted, however. The Maya knew perfectly well that the pigs were the town's main weed and garbage control and thus a force for good sanitation rather than a problem. They also knew that arguing with the public health personnel was a waste of time. Conversely, the deadly serious business of assigning acreage for families to cultivate was pursued with animated interest, and the final decisions (reached after much solid debate) were carried out to the letter.

Ejido decisions are merely the tip of the iceberg. The vast majority of the really important discussions take place informally, within families and between neighbors. And it is in these that the philosophy of interaction with nonhuman beings comes through most clearly.

When Noemy Chan saw her children playing by switching butterflies out of the air and killing them, she made them eat the butterflies. One does not kill except for food.

When a group of farmers weeding a field discovered a nest of rabbits, they carefully covered the babies up again and left them to live—though the rabbits were a major pest (see Anderson and Medina Tzuc 2004).

When anyone found an orphan deer or peccary or other large mammal, or killed a mother for food and discovered orphaned young ones, he or she would normally bring the young ones home and raise them.

Maya love pets, and most homes have several; pets are well fed as long as the family has anything.

Killing birds with slingshots—a typical amusement of children all over Mexico—was virtually confined to elimination of genuine pests. Moreover, when my conservationist friends Arturo Bayona and Adriana de Castro came to Chunhuhub and made a strong case against slingshots, the local children abandoned the toys without protest. They had few slingshots before, and they now have fewer.

When Maya picked up ants or bugs to show me, they would tenderly replace them—unhurt—back where they were found. Sometimes they were careful to put them back headed in the direction they were going when seized.

Maya also care deeply about their bees, viewing them as real companions, not mere mindless honey-makers (see extended discussion in Anderson and Medina Tzuc 2004).

When I asked my field assistants about the uses of plants, they often included the uses made of the plants by woodpeckers, warblers, or other animals of no immediate use to the human species.

Bird and insect damage is tolerated: "They have to eat, too." Betty Faust records, from Campeche, an even more dramatic observation: when she slapped at mosquitoes and complained about them, "I was reprimanded with the reminder that biting people is, after all, the 'work' of mosquitoes" (Faust 1998:110). God assigns work to every animal, and biting people is the proper role of the mosquito. I must admit I have not observed such forbearance in Chunhuhub in regard to mosquitoes, but I have seen it in the case of leafcutter ants. When these animals devastated one end of Don Antonio Azueta's orchard, he told me that some people burn or even dynamite the nests, but he had no intention of doing so. The ants were part of the place; they had a right to live. I have seen this tolerance of leafcutter and other ants in many cases. Swarms of army ants often invade Maya houses and go through them thoroughly, eating every small insect they find. The Maya see this as a useful housecleaning service. More than once I have sat with Don Felix and others, watching and commenting as the army ants milled around us and systematically explored the house we were occupying. (By contrast, even the most ecologically minded outsiders often reach for the insect spray in such cases—endangering their own health, for the amount of insecticide needed to stop an ant army is enough to be seriously dangerous to humans.)

Countless small measures involve people religiously in the landscape and its plants and animals; hanging a jawbone in a calabash tree, to make it bear more fruit, is one small example.

All these and other small matters, taken together, reveal a very different view of nature from that common among Hispanic Mexicans or contemporary United States farmers (Anderson 1996a).[2] In some Maya communities, often those more acculturated to Hispanic Mexican ways, slingshots are in constant use, domestic animals are abused, and pets are maltreated. A less cruel but quite callous farming method exists in the contemporary United States; rural animal treatment is often coolly instrumental—animals are treated as things, not as lives. One need think only of battery farming of chickens and pigs, and even of puppies, thousands of whom are reared under circumstances so psychologically damaging that they cannot become companion dogs. Such farming of pigs and poultry exists in a limited way around Merida, but it is totally incompatible with traditional Maya agriculture and values.

The Maya attitude extends to hunting. Animals are killed for food, but only for immediate necessity. Killing too many is punished by the Yuntsiloob, or by God (Anderson and Medina Tzuc 2004; Llanes Pasos 1993). Individuals hunt for food for their families, and sell excess meat, but specifically market-oriented hunting does not exist in the area—except among a few despised moral outcasts. Individuals kill very few animals and feel a deep and intense moral charge not to kill many; this is backed up by a real fear of supernatural sanctions. All this, unfortunately, does not necessarily stop overhunting (Anderson and Medina Tzuc 2004). There are simply too many people and too few animals in Chunhuhub. Everyone taking just one animal would wipe out the larger game—and, indeed, large game is now exceedingly rare in the area. The neighboring town of Polyuc is named after the brocket deer (yuk), now absent from this entire part of Quintana Roo. The white-lipped peccary is also gone. White-tailed deer, collared peccaries, ocellated turkeys, curassows, and guans are almost gone, though a few still exist in remote parts of the ejido.

The strength of conservation morality varies in Quintana Roo; it is sadly weakened in areas near Felipe Carrillo Puerto (as I have seen, but I rely more on Jeffrey Jorgensen's excellent studies of X-Hazil; Jorgensen 1994, 1998). It is gone, or nearly gone, in areas near Cancun and Chetumal.

It is still alive, but challenged by young poachers, in communities like Chunhuhub and Manuel Avila Camacho (Anderson and Medina Tzuc 2004). It is alive and flourishing in Tres Garantias and Caobas, where wildlife conservation is aggressively and successfully pursued with the help of the government (Anderson and Medina Tzuc 2004; Jorgensen 1998). The most important variable is proximity to a town, with its corrupting influences and its markets for meat. However, cooperation with the governmental conservation and management plan, the Plan Forestal, is also significant; this plan provides expert support for conservationist attitudes and programs.

Plants, too, are matters of concern, though of a less tender sort. Milpas are not only multiple cropped; the wild vegetation that grows up with the maize is usually slashed down, not uprooted. This allows it to regrow rapidly after abandonment. It also allows the farmers to spot valuable plants and spare them. Edible roots and shoots, medicinal plants (such as *kaambajauj, Dorstenia contrayerva*), and other useful plants are allowed to grow. Herbicide would be a disaster in such a situation, and the farmers know it; they sometimes use insecticides, but so far have resisted herbicide use. Spraying of roadsides by highway maintenance crews leads to annoyed comments that the herbs are rendered unsafe to eat or use.

A milpa, once "abandoned," is still managed and cropped for game and plants. Farmers continue to harvest surviving crops, as well as plants that were deliberately left intact or deliberately coppiced instead of being cut to the base. Hunters seek animals in the places they are known to prefer, such as the vicinity of the surviving crops: bean vines lure deer, fruiting squash vines lure pacas, small seeds attract quail and doves. Long-yielding milpa crops continue to bear after the milpa has long been left to regrow forest.

The result is an environment finely packed with diverse species, subsystems, and activities. Such a fine-grained environment has many virtues beyond its diversity. Among other things, it improves work efficiency, by allowing individuals to do many things on one trip. A visit to the milpa may involve cultivating growing crops, harvesting already-grown ones, gathering medicinal herbs or wild fruit, gathering firewood, engaging in incidental and opportunistic hunting, spotting bee trees for future reference, keeping trails open, monitoring expected and unexpected changes in the environment, getting new seeds from a neighbor, and exchanging labor or news, to say nothing of enjoying recreation.

Woodsmen, travelers, and others continue to harvest old-growth forests that appear "virgin." They comb these unobtrusively but regularly, seeking chicle, wild fruit, medicinal plants, honey, and game. Persons constructing houses or sheds take posts, beams, poles, vines, and bast for tying.

Significantly, there is no Maya word for "weed," in the sense of a plant that is purely a pest. (*Loob*, however, does mean a plant that is not notably useful. It has a focal reference to a small tree, *Eugenia mayana*, that is proverbial for its lack of utilitarian benefit. Yet, even this tree is used as a cough medicine and is admired for its attractive appearance.) The weed concept has come recently, along with the Spanish word for it (*maleza*). *Plaga*, "pest," has also entered the vocabulary, to refer to insect pests—since Maya has no good equivalent, though evaluative words like *k'as*, "bad," are applied freely. The people of Chunhuhub shoot coatis and other pests that take their corn, but only if they plan to eat them.

Apparently, this is a case in which economic use led to conservation. But, also, when a landscape is saved for any reason, people find uses for all that is saved, and thus sustainable management feeds back on itself. These and countless other examples of conservation are one manifestation of a wider management strategy found throughout the rural parts of the peninsula. (It has been described in detail from other communities: Faust 1998, 2001; Ramirez Barajas and Torrescano Valle 2000; Terán and Rasmussen 1994. On intensification and conservation in nearby areas, see also Primack et al. 1998; Remmers and de Koeijer 1992; Shriar 2001).

The Maya ideology for using the landscape was based on extremely extensive knowledge of the resource base. The Maya value and manage it for about a thousand plant species and perhaps a hundred animal ones. Many of these have multiple uses. Not only food and fiber, but aesthetic and ceremonial uses are extremely important. Ja'abin, for instance, is valued for its hard wood, which is important both for construction and for firewood; for its medicinal bark; for its beautiful flowers; and for its fresh leaves, which are ritually necessary for traditional altars. Trees were traditionally managed not only for direct benefits, but for wildlife, windbreak, and other functions (Remmers and de Koeijer 1992). Knowledge of the particular management needs of each important species is equally extensive. An enormous database lies behind the management strategies.

Another valuable effect of multiple use is that it makes people feel responsible for the total environment. The landscape is not just a setting for production of oranges or cows. It is a valuable system in itself. When two people meet, they often exchange information about the state of some part of the forest or the condition of a trail: "the old trail to so-and-so is getting grown over; somebody should clear it." It is a courtesy, observed by most older woodspeople, to cut encroaching vegetation as one takes a trail, so that the next person will find the path still open.

Under these circumstances, it would seem almost necessary for the overall ideology of management to be conservation oriented. The whole forest is a productive unit—not just something in the way, and not just something to be mined for one or two commodities. Moreover, it would also seem inevitable that management becomes an intensely emotional issue. The Maya love the flowers, birds, and wildlife of the forest.

It is thus not surprising to find that management is religiously constructed. It rests on emotion, security, community, mutual responsibility, and other matters that are inseparably involved with religion in all human societies except contemporary dominant ones. In Max Weber's terms (Weber 1951), the Maya are still enchanted: they still live a life in which religion is intensely bound up in daily experience and interaction, including interaction with plants and animals. Weber used the word "enchanted" in its literal sense; Maya traditional ceremony is indeed based on chants (Montemayor 1995).

There is no "nature" in Maya. Maya speakers must borrow the self-consciously learned Spanish word *naturaleza*. The Maya distinguish between the settled world—*kaaj*, "community" and kool, "fields"—and the *k'aax* ("forest"). These interpenetrate. K'aax soon returns to reoccupy abandoned fields, and even invades the town. All k'aax is merely regrown fields—for the Maya are aware that most or all of the land has been cultivated, and any farmer can relate the cultivation history of plots that appear "virgin forest" to the outsider. The k'aax is the habitat of things that are *baalche'*, "things of the trees," as opposed to *'akbij*, "tame," or *'alakbij*, "household reared," a distinction that separates nature from nurture. But there is never a contrast between "man" and "nature," *hombre* and naturaleza. *Winikoob* move easily in both wild and tame biorealms. The supernatural beings that are so close to humans are more firmly fixed in the environment; they live in trees, caves, ruins. They personalize the wild as well as the town.

Younger Maya who have lost this elder faith are becoming more and more prone to separate man and nature, but in the rural Zona Maya they never seem to take on the extreme views of traditional Hispanic Mexicans, who see nature purely as a threatening presence to be dominated or destroyed. For Hispanic Mexicans, even acknowledged benefits such as game and timber were until recently eliminated as fast as possible, to be replaced by livestock and plantations (as Lenkersdorf relates; see also Alvarez del Toro 1985). Multistranded involvement with the environment predicts the level of concern, and therefore it is most pronounced in the more rural and subsistence-oriented individuals, and especially in the older generation. The younger and more outside-oriented are, in Max Weber's terms, relatively "disenchanted." However, literally everyone I have questioned in Chunhuhub and neighboring villages expresses a love, respect, and concern for the environment, wild or tamed.

Insofar as I have observed actual behavior, it parallels the stated concern. There is, of course, some slippage. People everywhere are prone to talk better than they act. In Chunhuhub, poaching, overkill of game, and illegal logging are far from rare. However, the area has been spared the wholesale destruction that one sees in some other areas (Faust 1998, 2001; Shriar 2001; Terán and Rasmussen 1998).

Here as elsewhere, my work provides a close comparison to the similar investigations of Atran (Atran 1993, 1999; Atran et al. 1999), Faust (1998), and Terán and Rasmussen (1994; Terán, Rasmussen, and Cauich 1998)—to say nothing of Barbara Kingsolver, who has described the ecological delights of Calakmul, Campeche, in a recent book (2002). Terán and Rasmussen studied Xocen, a community I know well; it is not far from Chunhuhub—just across the border into Yucatán state. It has also been studied by my friends and sometime coworkers, Luz Elena Acosta (Acosta et al. 1998) and Bruce Love (Love and Peraza Castillo 1984). Faust studied Pich, a village in Campeche that has lost much of its Maya culture, including indigenous conservation attitudes. Resources there have come to be treated as wasting assets. Faust's older informants decry this and lament the loss of traditional conservationist values. My town of Chunhuhub is somewhere in the middle—much more outward looking than Xocen, but more traditional than Pich.

Atran and his group (Atran 1993, 1999; Atran et al. 1999, 2001) have done particularly important work on the difference between Itzaj (linguistically close to the Yucatec), Hispanic, and other Mayan farmers in Guatemala. The Itzaj management system—essentially the same as that of Quintana Roo's Yucatec (who are, in fact, mentioned in Atran et al. 1999)—has been learned by many Hispanic immigrants (as it has in Chunhuhub and elsewhere in the Mexican Yucatán), but Maya immigrants are resistant, feeling that, as Maya, they already know enough. They are thus mismanaging the forest. I have heard of similar problems in Campeche and Chiapas.

A wider context is provided by Bray et al. (2003). Their article title says it all: "Mexico's Community-Managed Forests as a Global Model for Sustainable Landscapes." Much of the article is (unsurprisingly) about Quintana Roo and the Plan Forestal villages. The authors agree, broadly, with my findings, and provide valuable technical details.

The poorest and most marginal of subsistence cultivators still manage to be, by sacred tradition, conservationists. They are interested strictly in "wise use" or "sustainable exploitation," not in preservation for its own sake, but they know that unwise use meant death in the old days. For one thing, they are used to saving seed. Seed is tempting during hungry times, but inviolate unless even wild foods are exhausted and frank starvation is at hand. They are also aware of the need to manage breeding stock of domestic animals.

The Maya of Quintana Roo know the land intimately and in incredible detail. This is a pragmatic knowledge; it is primarily about making a living. They use almost all important or obvious species in the environment, as well as countless obscure ones. They draw on local plants and animals for everything: food, fiber, fuel, toys, aesthetic delights, sacred rituals.

This constant intimate interaction, inevitably, involves the whole person. A Maya woodsman or mother is totally involved with these critically important plants and animals, rocks and soils.

Also inevitably, since humans are creatures of emotion as well as cognition (Milton 2002), they represent and construct landscape in emotional and spiritual ways (Tuan 1990). The Maya have taken this process very far indeed. The fields and woods are sources of symbols, meanings, and emotional expressions. They are involved in religion and ceremony, story, and reminiscence.

This involvement, in turn, feeds back on actual use. It leads to—and is the expression of—a strongly conservationist ethic. This ethic would not satisfy those modern conservation biologists who yearn for total preservation (Kay and Simmons 2001), but it has served to preserve the Maya forests for thousands of years. Even in the late Classic period, when much of southwest Quintana Roo must have been a vast cornfield not unlike modern Illinois, there must have been pockets of forest and brush, and these must have been duly appreciated.

In many areas of life, conservation has declined. For instance, few trees are left alive in larger fields today, after burning. Careful clearing around valuable wild trees (sapotes, wayas, and others) is less common than it was ten years ago. Fast-regrowing trees are less often coppiced, more often simply cut at the root and thus killed. (All these are changes I have recorded in several hundred miles of trail and field walking over twelve years.) The spirits of the forest are no longer performing their guardian function as well as they once did. Even so, traditional ideology is strong, and total forest cover is about the same in 2004 as in 1990. Slash-and-burn fields grow back, often to forest rich in mahogany and other valuable and encouraged or planted trees.

By contrast, "modern" agriculture in the area tends to feature total destruction of the forest to replace it with monocropping. In Chunhuhub, oranges and (very locally) watermelons are monocropped. This involves creating huge, pure stands of one species, and viewing all other species as pests, or at best invaders. The Maya of the Chunhuhub area are broadly tolerant of such invaders, and continue to use orchard weeds and animals for food, medicine, and other values. However, a different attitude certainly comes with monocropping. Suddenly, an invidious distinction is made between one species and all the rest. Not only is one working for only one crop; the monocrop system encourages the few plants that even Maya regard as purely pernicious—mostly introduced weedy Eurasian and African grasses such as Johnson grass (*Sorghum halapense*, called *zacate yonson* by Spanish and Maya in Mexico).

Monoculture is basically the child of the plantation, at least in tropical America. It arose when rich, foreign planters came with the goal of making maximum money in minimum time. They had no knowledge of, or vested interest in, the habitats they ravaged. Large-scale monocropping has persisted not because of economics but because of the plantation heritage and, more recently, because it is easy for states to control, and thus is pleasing to bureaucrats (Scott 1998). Monocropping may be economic in some tropical situations, but usually it has been economically problematical.

Shifts from production for use to production for sale often affect resource management decisions in this area (as in others). Often, production for sale leads to a more rapid and drastic simplification of the local ecosystem. This can be observed in most of the tropics. The traditional Maya cultivation system not only conserved biodiversity; it probably enhanced it (Fedick 1996; Gómez-Pompa et al. 2003). By contrast, production for sale has resulted in monocrop cultivation of henequen, oranges, cattle, and other commodities. In the Yucatán Peninsula, at least, this not a result of the capitalist market nexus, but of cultural bias. (In some other parts of Central America, the penetration of the cash economy seems more directly causative of similar changes; see Schwartz 1990; Stonich 1993.) The Maya prefer to maximize diversity, and some of them have very successfully adapted their traditional mixed-species dooryard gardens and subsistence farms to the market economy. In fact, in Chunhuhub, mixed-species market gardening and orchard keeping pay well when seriously attempted. Monocropping comes from a different cultural world: the world of Mexican government planners and large-scale agriculturalists. These developers are imbued with the "modern" notion that successful agriculture must be mechanized, chemical-intensive, monocrop cultivation of salable commodities. This is an entrenched belief, not a proven fact of economics; all the evidence runs the other way. However, the belief in "modern" agriculture is held with literally religious fervor by many development personnel, in Mexico and elsewhere, as I know from many a long conversation. It goes with the belief that anything the "peasants" do, especially if they are "Indians" to boot, is necessarily superstitious, backward, and foolish—not only uneconomic but downright offensive and embarrassing. (For detailed accounts of this process, and the irrationality of "modern" agriculture in the Yucatán

Peninsula, see Faust 1998; Faust, Anderson, and Frazier 2004; Hostetler 1996.)

Thus, instead of individual rational choice predicting the economic activities of the Yucatán Peninsula, we have two cultural systems in action. Neither is perfect; both could be much improved by more rational technology. However, one of them has the benefit of five thousand years of experience in this challenging environment. The other has no such history; not only has it been tried for only a bit more than a century, but also its bearers have failed to learn much from experience. The decline and extinction of economies based on monocrop agriculture—successively on cotton, sugarcane, henequen, rice, sesame, and other commodities—have taught the development agents rather little. The story is sobering to anyone, but perhaps especially to those who labor under the illusion that direct economic "rationality"—specifically, in agriculture, the quest to maximize profits or calories—is the sole important determinant of human behavior. The real determinants are more complex.

James Scott's brilliant book, *Seeing Like a State* (1998), shows at great length that the modernist view is a true belief system—based not on evidence but on prejudice—and that it is economically irrational but highly predictable from the power relations involved. He points out that this notion was adopted for cultural and political reasons, not for economic ones. Large-scale monocrop agriculture looks clean and neat; it is known in the United States as "clean farming." By contrast, even sympathetic observers are wont to describe Maya gardens as "messy" (as one anthropologist did, to me, a few days before I wrote this paragraph). Little do such observers realize that every plant has to be carefully fitted into its microhabitat—the one spot in the garden that has enough of the right kind of soil, or is near enough the water supply, to be optimal for that plant. Straight rows do not survive, except in the few areas where extensive tracts of deep, level, fertile soil has accumulated in low areas at the foot of the hills.

Government planners like monocrop cultivation because it is easy: "clear everything and plant 100 ha of henequen" is an easy way to plan for 100 ha, whereas the traditional Maya system requires that the farmer must know every inch of the hundred-hectare tract and manage it plant by plant. Perhaps more to the point, industrial agriculture is easily controlled. Politicians like it because it is centralized, being usually directed by a tiny elite—government planners, big landlords, or giant agribusiness firms. Politicians can thus extract taxes easily (Scott 1998) and also develop close, mutually beneficial relations with the large landlords and firms. Among other things, politicians in Mexico and the United States routinely exchange huge subsidies to such producers for huge campaign contributions from them.

Also, as Ueli Hostetler says of government plans in southeast Mexico, "people know that the money is spent anyway, not least of which because this allows involved bureaucrats and politicians to get their share" (Hostetler 1996:197). Here and in the discussion leading up to this conclusion, Hostettler, writing on eastern Quintana Roo, remarkably parallels Scott's thinking. Hostetler's community is culturally very close to Chunhuhub.

The "modern" agricultural system is probably never economic in the long run, since it is notoriously devastating to the landscape—soil mining, polluting, water wasting. Smallholders almost always husband their resources better and show higher profits per acre over time, a point definitively established by the life work of the late Robert Netting (Netting 1993).

Large-scale mechanized agriculture pays when labor is scarce and expensive; this is why it was developed in the first place (Hayami and Ruttan 1985). It is also useful in certain habitats where only one or two economically valuable plants or animals are really well adapted. It is thus appropriate for wheat farming in Montana, where both labor and alternative crops are hard to find and where industrialized wheat farming does not devastate the land. It is reasonably appropriate for maize farming in Iowa or tomato farming in California; better systems could be imagined, but given current facts of land ownership and labor availability, the system works.

It is definitely not appropriate for land-poor, labor-rich Mexico, or other tropical lands. Modern agriculture in such cases survives only because the government pours into it vast subsidies, not only directly (in the form of cash payments), but also through flood control, soil conservation work, technical research and extension, regulation of pesticide and fertilizer use, infrastructure development, and much more. Mexico is not in a position to succeed indefinitely at this.

Pesticide regulation, for instance, is difficult enough at best in a developing country; but the chemical companies render it more so by providing many incentives to government officials to recommend heavy use of these chemicals. The result in many areas has been an overuse of pesticides, which not only damages the soil and causes pests to evolve resistance, but also endangers human life (Simon 1997; Wright 1990). Flood control and water management, so necessary to success when one is using modern high-line crop varieties, has barely begun in western Quintana Roo. Thus, to cheap land and expensive labor, we must add a strong and supportive government as a necessary "factor of production" for modern technology to flourish in agriculture.

In the peninsula, short-term successes, most notably the famous henequen boom of the late nineteenth and early twentieth centuries, have blinded Mexican and international planners to the long-term unsustainability of monocrop agriculture. In Yucatán, the henequen estates are abandoned, the sugar haciendas long gone, the cotton and chile and tomato fields mostly gone back to brush. At present, oranges are successful, but only because they are grown by smallholders as part of a semitraditional Maya mixed system.

Another reason for planners' love of "modern" agriculture is that it is entirely integrated into the cash market. Subsistence agriculture produces benefits that do not show up on balance sheets. This looks bad to planners (my personal experience certainly confirms Scott's general point in this regard). In fact, a Maya family living by subsistence production in western Quintana Roo is producing and enjoying goods and services worth between $2,000 and $3,000 (US) on the open market. Moreover, many of these farmers produce for the market, but report no income because they barter, or because the amounts are so small that the government does not trouble to collect the statistics. If all this were costed into the government accounts, Maya small-scale agriculture would show its formidable importance in the Yucatán economy.

Insecticides provide another case in point. Reasonable enough in some cases (when short-lived pesticides are applied on a local basis and at safe doses), pesticide use has become as pathological in Mexico as in many other areas. Applications that are many times the safe dose are spread over vast areas. The result is that insect-eating creatures of all kinds are killed—from birds to parasitoid wasps. Resistant strains of pest insects soon arise and breed without control. Not only do they kill the delicate, heavily sprayed commercial crops; they escape into surrounding fields and play havoc with traditional Maya crops that had previously withstood pest attacks (see Faust 1998; Wright 1990). A Maya field is a constant battleground in which insect predators routinely devastate pest populations; once these predators are gone, even the toughest Maya crop is vulnerable.

The problems of misplaced modernization in the tropics have been amply discussed in the literature (Faust 1998; Gonzalez 2001; Painter and Durham 1995; Sponsel et al. 1996). What has not been so well discussed, partly because so many "critical" writers have very little training in agriculture (and perhaps more interest in "critique" than in solving the problem), is the appalling opportunity cost of this sort of misapplied modernization. The Sosa farm, of Naranjal Poniente's mahogany management (Anderson 2003), of Tres Garantias's integrated planning (Anderson and Medina Tzuc 2004), of Chunhuhub's mixed orchards, and—for that matter—of the successful cattle raising and dairying of the natural grasslands of Quintana Roo give a sense of the problem. All these very different schemes have one thing in common: they produce an enormous amount of use-value, sustainably, on very little land, and that land none of the best.

A little imagination, and some experience with experiments in other parts of the tropics, opens up far more possibilities. I have seen cherimoyas sold for $4.00 apiece in Los Angeles, and I have seen them fed to the pigs in Chunhuhub. Similarly, I have seen American cabinetmakers and woodworkers pay exorbitant prices for woods that are burned or, at best, used for rough construction in Chunhuhub. All it would take is some minimal enterprise in marketing to bring the resource to the top-dollar buyer. That minimal enterprise has been effectively stifled in Quintana Roo by the twin blankets of bureaucratic impediments to private business and bureaucratic commitment to poorly planned public projects.

In particular, marketing has almost never been taken seriously in government plans in the Chunhuhub area. I have personal experience with several development plans, made with the best of intentions and administered by genuinely competent and dedicated people, that failed because no one thought of marketing issues. Vegetable production was undertaken miles from a road, honey production was started without quality control demanded by export markets, woodcutting went on without taking into account the need of temperate-zone woodworkers for very carefully seasoned wood, fruit raising for juice concentrate was undertaken when the only market was for fresh product, rabbit raising was tried in an area where no one eats rabbits (and chickens are cheaper anyway), and ecotourism was developed without any thought of advertising in tourist-sending countries. Thomas Dichter (2003) provides a whole collection of similar stories from around the world, including a rabbit-raising story from Ghana that is both truly dark humor and very close to Quintana Roo experience. Nor are bureaucratic planners the only ones who make this mistake. The Sosas' visionary farm gave way to a beer hall for the sad reason that Chunhuhub's consumers would rather drink beer than eat kohlrabi and beets.

Chunhuhub remained a town based on mixed traditional agriculture until the rise of citrus as a Maya cash crop. Citriculture expanded in Yucatán state in the 1960s, and by the 1980s was dominant and enormously successful in Oxkutzcab and elsewhere (Morales Valderrama 1987). By the end of the 1980s, it was expanding rapidly into Quintana Roo. It proved far less damaging to the environment than most monocrop schemes. Citrus trees, especially oranges, are well adapted to the climate and the thin, loose soil. Being deep-rooted, shady tree crops, they tend to preserve the soil from erosion.

Moreover, the government, in a rare moment of genuine inspiration, took to clearing narrow strips earmarked for citrus along the highways, and bringing irrigation to these. The land was then leased to local farmers—through the ejido, if the land was ejido land; directly, otherwise. As with ejido land, farmers could lease up to four hectares, depending on family size. These "parcelas" proved popular and became the main cash sources for countless families. The highways are thus now lined, for miles, with groves of flourishing citrus. This proved to be an ideal way to grow citrus, because the trees were easily accessible from the nucleated Maya villages. Also important was an unplanned side benefit: the success of "citrus corridors" in saving the roadsides from casual low-payoff cultivation, pasture clearing, random burning, and the other pernicious highway-following activities that have made roadways so dreaded by environmentalists throughout the tropics.

However, oranges are not very profitable. They do not bear as well as they do in more reliable climates. They glut the market during harvest time. For large-scale production of juice concentrate, they are undercut by competition from Brazil, where production is more rationalized and costs are lower (information from interviews with government planners). In Quintana Roo, any adverse event, from Hurricane Roxane to a fall in world orange prices, devastates the orange economy. Other citrus trees are still less profitable, being salable only as fresh fruit to the market. It can never be more than one part of a larger and more complex economic adjustment.

In some settings, modern industrial monocrop agriculture produces more economic revenue than traditional diversified farming. In northwest Mexico, for instance, highly profitable vegetable and cattle raising has arisen on desert lands. The social consequences of this farming, as well as the long-term environmental consequences, are unfortunate (see, e.g., Wright 1990), but at least there are some benefits to some people. In Quintana Roo, except for the limited success of the orange industry and the success of cattle on savannah lands, even short-term benefits have failed to emerge.

The traditional mixed, fine-grained, richly diverse cropping system not only fits the local ecology; it fits the local socioeconomic situation. Labor is cheap, abundant, and highly skilled. Thus a labor-based, knowledge-based system makes sense. At the other extreme, mechanized agriculture is not only an ecological disaster; it also replaces skilled labor with a machine that cannot plan for microhabitats or know which weeds to spare, and that constantly needs expensive and hard-to-get replacement parts. The displaced workers drift to the cities, where they find themselves "unskilled." Their profound knowledge base does them no good. In fact, it brands them as hicks. It is a lose-lose situation. Chunhuhub's one tractor is useful (when it works), and pumps for irrigation water are truly a blessing,[3] but full mechanization would not be to anyone's benefit.

In the last analysis, the core point is that the Maya manage for maximum diversity: Diversity of species, diversity of landscapes (including the fine-grained pattern of regrowing milpas), and diversity of uses of everything the environment offers. As the world enters the twenty-first century, the survival of tropical peoples depends on such diversity and sustainability. Within a very few years, all the world's valuable tropical woods will be cut, except in a few areas (hopefully including interior Quintana Roo) where sustainable forestry will allow a small but continuing production. At the same time, virtually all large and edible animals will be extinct. Within a very few years more, all tropical soils will be degraded, except in those few areas with exceptionally rich land and exceptionally careful management.

Much of the tropics will resemble the largest-scale clearings of the Chunhuhub area: worthless scrub, hopeless for even the minimum cultivation, and hopeless for reforestation in the foreseeable future. This is not, alas, merely a cynical forecast of the future. It is a realistic view of the present. Outside of Amazonia and a few remnant areas of Central America and interior Southeast Asia, it is already fact for much of the world's tropics.

Assuming that Quintana Roo is lucky—a contingency that waits on further serious measures to stop rural population growth—the Maya will have to build on their existing system of multiple use. There is, realistically, no future in monocropping or other "modern" agriculture. There is also a need for some kind of intensification. Traditional agriculture is not viable by itself. First, the population density is already too great (Terán and Rasmussen 1994). Second, needs for cash are increasing in the modern world, and the traditional system produced almost exclusively for direct consumption. Outside of a small amount of money derived from honey, chicle, and maize, cash tended to come from unsustainable logging and hunting. In future, logging can be managed sustainably (but with reduced total profits, obviously), but market hunting cannot be; even subsistence-level hunting is no longer possible without major controls.

Most wonderful of the projects in central Quintana Roo are the efforts of Arturo Bayona and Adriana de Castro. Their "Casa Econciencia" in Felipe Carrillo Puerto has been the center for environmental activities for the region for many years now. They have worked especially with children and in education (see Ancona y Rivera et al. 1995), on projects ranging from anti-slingshot campaigns to major initiatives in guidance, documentation, and education. They have done it independently, on a shoestring.

Education is indeed the key, in the long run. Ricardo Godoy has recently found that a year of education, for country people in Honduras and Bolivia, reduces the area of forest cut by 7 to 17 percent (Godoy 1999). This is more because the people learn to make a living without subsistence farming than because they learn how to manage resources. Rural resource management is, alas, rare in the world's curricula. One can only imagine how much better things can be if the plans and dreams of Bayona and de Castro are implemented, if their textbook (Ancona y Rivera et al. 1995) and others like it are used, if rural people learn from the formal establishment what they used to learn informally. Of all needs, the most desperate and the most unrealized is for environmental education for the world's children. It is sad that the formal system is coming to substitute for the informal one, but that is reality. One can only hope that the informal system will persist and complement a newly responsible formal system.

What has turned things around in Quintana Roo is the presence of a handful of responsible people with adequate biological training, in a wider setting of changing consciousness. The handful—ranging from professional biologists like Albert Ehnis Duhne to educators like Bayona and de Castro—has been critical in actually working out plans and bringing them to the public. Without them, nothing could have been done; there would not have been enough expertise to allow construction of successful plans, either for managing resources or for teaching the public about them. Without the wider context of changing views, no one would have given them a forum or implemented their ideas.

The traditional solution to demographic pressure has been the export of people to daughter colonies. This will soon be impossible, as the land fills up. Export of people to the cities, and to other areas of Mexico and the United States, has become the standard coping mechanism today. It is inadequate; too many people remain on the land, and the jobs found by the out-migrants are too often dead-end and dismal (Kearney 1996).

Mexico is also exhausting its cultural-institutional base. Traditional concepts of conservation, management, and cooperation were sustained by a religion and worldview that are now increasingly modified by contact with other views (ranging from urban rationalism to fundamentalist Protestantism). New ways of convincing people to conserve resources are needed. Fortunately, they are coming, whether through urban media, local educators, or political discourse.

The Maya system is based on three mutually supporting realms of action:

Technology. The traditional system uses almost every prominent species of plant and animal for some purpose. The technology is knowledge-intensive. People know every conceivable use for every conceivably useful item in the forest—and "uses" include wildlife food and shelter. I recorded about 1,300 plant names, covering some 700 species; 352 species are used for medicine, more than 150 for food, around 180 as planted ornamentals. Local knowledge extends to knowing every inch of one's milpa, including the hollows of deep soil, the moist spots, and the particular trees. By comparison with typical modern agriculture, Maya agriculture is far more *knowledge-intensive.*

Organization. So far, Maya agriculture is organized at the ejido level and managed by the community and by extended families. By comparison with the competitive individualism of modern agriculture, it is *collective.* Ejidos are legal corporations, and families act as corporations. Corporate management works well insofar as it is horizontally and vertically integrated.

Ideology. As we have seen, Maya ideology relating to environment use privileges the *long term.* Also, it must calculate benefits for the whole managing entity, so the calculus is much more *wide-flung* than is typical in modernized farming. Moreover, the intricate religion, which has survived in a Christianized state and remains a living force, involves individuals emotionally in the land and creates internal self-regulation—a conscience. Ideology is generally shared, not a tool used by elites to manipulate the masses. (Elite manipulative ideology does, of course, exist in Quintana Roo, but the Maya of Chunhuhub are far from the elite levels at which it operates.) Ideology becomes part of the knowledge system that is necessary to manage the forests and fields. It is part of a feedback process, a polylogue, not a one-way or top-down imposition of voice. Thus, the three components of the system are in constant interplay. It is not the one-way determination of social superstructure by economic foundation that at least some—perhaps naive—readings of Marx suggest (one can read Marx as closer to the foregoing position; v. Marx 1909 [1867]).

Maya environmental philosophy is related to that of other Native North American groups. Personalizing the environment and viewing it with emotional and spiritual concern is common throughout the continent (cf. Anderson 1996a; Berkes 1999; Callicott 1994; Hultkranz 1967). The people of Chunhuhub do not attribute souls or spirits to all natural phenomena, as Northwest Coast peoples do, nor do they seek for visions or otherwise engage in many of the widespread North American religious activities that involve the environment. However, they are tied to the environment by many ritual activities and by obligations in ordinary life (such as the obligation to take no more game than is immediately necessary, a rule widely recorded in North America).

There are, thus, both culture-historical and functional reasons for Maya environmental philosophy to take the form it has. The Maya share a North American heritage. They also are involved with most local species, wild and tame, and with the whole landscape; they thus can hardly escape the need to manage the environment comprehensively and to feel some emotional involvement with it.

Social sanctions, including emotional and expressive sanctions in myth, ritual, and social criticism, do indeed exist in Maya resource management (Sharon Burton, unpublished research, 1990–1992; Hostettler 1996, 2002; Terán and Rasmussen 1994; Villa Rojas 1945, 1985). They do not work well today, because they are fading from people's minds, and because the world has been changed by rapidly expanding population and even more rapidly expanding resource exploitation. The consensus among investigators is that the rituals did once guide resource management quite well. They still have an effect, primarily by helping to bring people together to work and negotiate as a unified group, but also by instilling a strong conservation ideology (to be discussed in due course). Unfortunately, we will never know how effective the rituals were in the distant past.

The Yucatec Maya have accumulated, over thousands of years, a complex body of social and ritual activity centered on food production and consumption and on proper use of resources (Villa Rojas 1945, 1985; Thompson 1971). This presumably serves to transmit information about the system and to motivate individuals to act within it. Following the theories of Rappaport (1971, 1984) and Flannery (1972), one would expect that much of this information, as well as motivation, is encoded in religion and ritual; and indeed a significant portion of Yucatec ritual is directly associated with food procurement (Redfield 1941; Redfield and Villa Rojas 1934; Thompson 1971; Sosa 1985). Ritual regulation of resource use has been documented elsewhere in North America (Hunn 1982; Martin 1978; Morrell 1985; Nelson 1983; Ridington 1981; Swezey and Heizer 1977). Management may be implicit (cf. Rappaport 1984) or explicit. It may be exclusive if resource use limits are imposed on outsiders specifically, and inclusive if imposed on one's own group (Hunn 1982:20–23). In studying south China, I stressed the fact that the south Chinese rice agriculture system had flourished for thousands of years, supporting more and more people while actually increasing the fertility and extent of the paddies (Anderson 1988; Anderson and Anderson 1973). Robert Marks has recently come to a very different conclusion, pointing out that large animals and old-growth forests were sacrificed in the process, while population growth and urbanization eventually forced the densely settled areas to import food (Marks 1998). We both agree on the facts, and we are both right; I looked at the agricultural system as such, while Marks looked at the whole picture. This is not the "Rashomon effect" in action. We simply looked at different system levels.

One can look at a traditional biological system from two different perspectives. First, and perhaps most important at this stage of knowledge, we desperately need to understand such systems in their own terms (the "emic" view). The emic view is appropriate for anyone trying to understand a traditional system and understand how individuals think and behave in reference to it. Second, we need to look at them as sources of useful knowledge for the world at large, and thus to subject them to an external, independent evaluation based on cross-culturally applicable measures (the "etic" view).

Today, many scholars appear to believe that these ways of looking at a thought-system are mutually exclusive. There are those (largely anthropologists) who view any etic approach as an intolerable imposition of alien values and standards. Conversely, there are many (largely economic botanists and agronomists) who are only concerned with obviously and immediately useful knowledge, and who dismiss such matters as religion and even folk medicine—often referring to all such things, contemptuously, as "mere superstition." Both of these positions represent disturbingly narrow views. This is especially true if one's concern is with saving human lives on a worldwide scale.

The Quintana Roo Maya have managed to support a fairly large population with minimal damage to the environment, but, even so, conservation biologists cannot be expected to see them with a purely favorable eye. The fields of conservation and development need more serious dialogue between people holding various viewpoints. In such a dialogue, indigenous voices should be prominent.

If population rises, the traditional system almost inevitably becomes less and less sustainable. Practices that were sustainable, even ideal, at low population densities become unsupportable. To accommodate increased population, the traditional system must intensify. If this system is to remain at all sustainable, people must become more conservation-conscious—both to offset the increased direct pressures on the environment that result from more people using it, and to offset the indirect pressures caused by changes due to intensification. No system, not even the most conservationist, can possibly stand up under the progressive distortion of unchecked population growth. This is not to say that population growth is always and necessarily bad; population growth forces progressive change, and this change must include increasingly tight controls on environment use. This, of course, progressively distorts "traditional" culture; by definition, tradition is what does not change rapidly. Many traditional communities can respond rapidly and effectively to change (see, e.g., Gonzalez 2001), but many do not. Either there must be new and better ways of using the environment more efficiently, for which someone must plan, or the poor must face immiseration and suffering. The typical case in history, and certainly in most of the tropics today, is that the poor lose out.

CHAPTER SIX

The Town of Chunhuhub

ON STILL NIGHTS, you can hear the pottery. It recorded, by some kind of echo resonance, the sounds around it. Just before dawn, if you visit the ancient sites, you can hear the voices of the ancient Maya, chanting and playing their musical instruments. Maybe some day, we can amplify this and listen, and find out what life was really like back then. . . .

So I am told. I have heard also that the aluxoob, the ancient Maya figurines, come to life at night and move around. You can get them to guard your field; find one, feed it some corn bread and pozole, and order it to stay there. It will throw rocks and make terrifying whistles if anyone comes to steal your harvest. There are other images that are filled with gold. Sometimes they and the gold mysteriously disappear.

I have not yet heard the pottery talking, nor seen an *alux* move around, but that, they say, is because I have not been out at night long enough.

I'm writing in a concrete-block shed, in Don Pastor Valdez's home compound in Chunhuhub. Outside, a tropical night grows more and more quiet. This is the end of January, so the temperature is only around 75°. By May, even the nighttime temperatures will be in the 90s.

Chunhuhub itself is a town with at least a 1,500-year history. Don Pastor, digging a pit in his backyard, dug through colonial potsherds (typically seventeenth-century Spanish-style ware) to simple utilitarian precontact material, and through that to a large stone structure with Classic Period pottery. Don Felix, cutting his 1996 milpa, exposed a striking Classic site, on a rich and fertile terrace overlooking a large and lush aguada. There was stonework on the cutbank dropping off into the aguada; the bank had been terraced and faced, as were many aguada banks in classical times (Scott Fedick, personal communication 1996, and my observations at Dr. Fedick's sites). In 2003, expanding his orchard, Don Felix exposed a ruin mound covered with beautifully painted sherds from the Classic Maya period.

Chunhuhub centers on a humble rock mound standing above a long, low trough—all that remains of a small Classic Maya site occupied some time between four hundred and eight hundred years after the Handsome Lord. The Handsome Lord, *Ki'ichkelem Yuum*, is the name of Jesus, in modern Maya. In older times, the name was that of the Maya maize god.

In Classic Maya days, the trough was a reservoir, some 50 meters long, holding water during Chunhuhub's fiery and merciless spring drought. The mound was once a small temple structure. From it, a priest could perhaps look over fields and see the much bigger mounds of Tampac, 5 kilometers down the road. More likely, tall trees hid everything beyond a few hundred meters.

Classic Maya city-states rose and fell. Chunhuhub, already a sizable town, must have changed hands many times and probably did not care much; it was far enough from the great power bases to maintain a fair degree of autonomy. Probably it was a tributary community to the great city whose ruins now compose the Ramonal site at Nueva Loria. Perhaps it fell under the power of the even greater metropolis of Coba, farther away in the opposite direction.

The nearest large site lies 20 km south. There, the great Holy Mountain at Altamirano, just west of Nueva Loria, rose high enough to achieve the dream of all ancient Maya cities: a mound that allowed the priests and lords to see above the forest canopy to the next city's sacred hills. Even today, more than a thousand years after its decline, Nueva Loria's great pyramid has a soaring and lyrical beauty that makes many of the local people who climb it believe that it still has some of its old divine power. The high pyramid of the 2-kilometer-square site at Altamirano fronts eastward onto a leveled area cornered by mounds, rising 20 meters, but on the west it drops 30 meters in one great majestic sloping line, for it is built up on the bank of the great Altamirano aguada. From the top, one looks over tall mounds breaking the forest canopy and over lakes and marshes where dwell egrets, wild Muscovy ducks, and king vultures.

At Margaritas—some 50 km to the south of Chunhuhub—there is a truly huge site. Clusters of pyramid mounds rise intermittently from an area of 100 square kilometers. The biggest is a vast raised platform, square, oriented east-west, with huge mounds on all four sides. On the south is a pyramid that rises about 10 meters from the high platform, then plunges 20 to 30 meters down to an aguada below. There is a cave in which one can hear roosters crowing; the believers think these are supernatural winds; the skeptics think they are echoes.

When the Spanish conquered the Yucatán Peninsula, they found Chunhuhub situated on the main road from the Puuc to Bacalar and Chetumal. Chunhuhub continued to flourish—a sleepy little town of milpas and fruit trees. Early colonial records reveal the usual small-town problems: failure to pay taxes, excessive distilling of raw rum. Modern records reveal some continuity in both regards.

Far more impressive than the humble Classical mound is Chunhuhub's great ruined colonial church. It was burned in the Caste War. Its hollow shell, a powerful, stark skeleton, stood unroofed and empty in the lonely forest for a century, and then watched silently as a town grew again under its shadow. In 1996, restoration began on the tall masonry structure. I climbed 15 meters of rickety scaffolding to see the angel's face carved in the keystone of the nave arch. The date above it read "Nov. 1785." Below it were the mysterious initials BDJOJ, and some broken bits of other letters. The message remains cryptic, sealed.

In the late colonial and early independent years of Mexico, Chunhuhub had no *haciendas*, *ranchos*, or *fincas* (Bracamonte y Sosa 1993:55). It was still a town of collectively managed fields. Chunhuhub survived and became the center when the local population was "reduced"—significant word—by being concentrated in a single settlement. Gates tells us that, by 1549, "four towns had been destroyed and the people driven in to Chunhuhub by the Franciscans, and the tax levy in '49 for its [conquistador] grantee was 300 tribute payers [i.e., more than 1,500 people], reduced in '79 to thirty" (Gates's notes to Landa 1973, p. 144). The 90 percent reduction was probably due largely to disease, but many of the people would unquestionably have been drifting back to their old towns and scattered hamlets, in true Quintana Roo Maya fashion.

In the Spanish Colonial *Relaciones* (1983), compiled around 1579 in answer to a government survey (yes, even then), a section on Chunhuhub (and Tabi) was compiled by one "Pero" (Pedro) García, resident in Mérida but "encomendero" of the two towns (de la Garza et al. 1983:161–67). Much of his description is essentially a rewording of the standard core of the *Relations*, thought to have been prepared by Gaspar Antonio Chi and collaborators (cf. Xiu C. 1986). He notes that one Baltasar Balam was chief of Chunhuhub at this time; Balam ("jaguar") is still a common surname in Quintana Roo. His (or Chi's) description of the area is still valid: a rocky region without surface water, whose inhabitants live on maize, beans, chiles, poultry, and game. They grew cotton. They resorted to wild roots during famine, as they still do. Only the "small dogs which they raise for food" (de la Garza et al. 1983:165) have changed their status; small dogs still abound but are not eaten. The people "were all idolaters and adored idols of wood, clay and stone" (p. 164).

Then as now, Chunhuhub was important as a way station on the main road from Mérida to Bacalar and Chetumal, a position that gave it strategic importance and brought it more than once into the edge of the light of history (Jones 1989, 1991). In the late seventeenth century, the population of the Bacalar district concentrated at Chunhuhub, making the area strategically even more important because of ongoing dealings with the still-independent Maya deep in the Peten around Lake Itzá (Jones 1989:63–71).

In 1847 Chunhuhub was destroyed in the Caste War. Virtually nothing is known of the western part of this area during that time. Eastern Yucatán state was soon reconquered, but Quintana Roo was independent until the beginning of the twentieth century, and the deep central interior was not brought under government control until the 1930s. Quinine probably had more to do with the reconquest than soldiers did; malaria was a formidable protector of the deep woods. General Bravo's reconquering expeditionary force must have passed close by Chunhuhub (Brannan and Joseph 1991, especially the article by Herman Konrad [1991:143–171]). Chicleros and perhaps loggers passed this way, the advance guard of capitalist development (Konrad 1991).

The Cruzob history of the Zona Maya gives it a special character; it is one of the few places in the New World where the Native peoples feel with considerable justification that they were not "conquered." The last battle was actually fought in 1933, in Dzula, the village directly east of Chunhuhub. The Maya did not officially win this battle, but they most certainly did not feel they lost. Peace finally came to the area in the 1930s. The great Yucatecan ethnologist Alfonso Villa Rojas, exploring in 1937, still felt very insecure (Villa Rojas 1945).

The Maya of the Cruzob lands have suffered enough defeat and oppression to empathize fully with the Maya of western Yucatán, but they feel they have maintained more independence and in some ways a more truly Mayan lifestyle. In fact, the situation is complex. The Maya of Yucatán state have maintained many traditions lost in Quintana Roo, as well as vice versa (Brown 1988 and personal communication with the author; Villa Rojas 1978). In some parts of Quintana Roo, Maya have sought to deny their heritage, and maintain that their ancestors were loyal—even when this is clearly not the case (Mossbrucker 1997). In contrast, most of the descendents of Cruzob in the Zona Maya—including Chunhuhub—are still proud and militant. On the other hand, they put a high value on tranquility, though some brag of past rebelliousness.

Chunhuhub was repopulated from Chunhuas, now a tiny village just west of Felipe Carrillo Puerto (the main city of central Quintana Roo). Several people moved from there to a hamlet called San Diego. They began to collect chicle and occasionally farm the area around 1933. In 1940, families escaping locust plagues moved on west and began to farm permanently in the Chunhuhub area. In 1941, Chunhuhub itself was resettled—officially getting permission from the municipio to found a new pueblo. Juan B. Xool Uitzil, a "Cristiano perdido, de quien sabe donde" (lost Christian from none knows where) to one of my friends, but actually from Chunhuas by way of the tiny hamlet of San Diego, found the site. He and others from San Diego and Chunhuas settled there in 1941 or 1942. His brothers and his relative Pedro Tun soon joined him. About fifteen families were there by the mid-1940s, coming from all over the Zona Maya and some from eastern Yucatán as well.

They found the ruined colonial church, still inspiring and impressive in decay, dominating an overgrown plaza where wild turkeys strutted in place of humans. Jaguars and tapirs abounded. Now, all these animals are pushed to the more remote fringes of the community lands. Chunhuhub has vastly overshadowed its original settlers' home villages, tiny hamlets 30 kilometers to the east.

The Xool family—notably the Xool Uitzil brothers Juan, Juan Bautista, and Mateo—was the major founding family, and the Xool lineage remains important. Mateo became the first person to dic in Chunhuhub, in 1948, and his is the oldest grave in the little *camposanto* under the huge old tree at the town's west border. Pedro Canul, Domenico Tun, Pablo Aban, and Silvestre Uc were among the other early settlers, and their families—especially the Tuns—remain important. In 1942 a teacher came, and in 1942 a school was built.

In those days, a goal of small rural towns was to become ejidos: collective landholding communities recognized as such by the state. In the 1930s, President Lázaro Cárdenas greatly expanded the rights of Mexican rural people to declare their communities to be ejidos. This gave them formal communal ownership of a sizable demarcated area of land. Heads of existing local families are formal ejido members. They run the ejidos by direct democracy, meeting and voting in asambleas. An ejido council manages day-to-day administration. Ejidos in Quintana Roo tend to be large; Chunhuhub eventually gained a square of land totaling 14,330 hectares.

From 1944 to 1950, the process of officially forming an ejido ground slowly on, and Dionisio Xool—my source for much of this history—was the first comisario. In 1956 a road reached Chunhuhub, and in 1968 the road head began moving south, soon connecting Chunhuhub with Chetumal by a good modern highway. This became a shortcut from Mérida to Chetumal, and thus a busy truck route, providing opportunities for lunch stands and truck repair shops along the road.

José Dolores Xool Aban is one of the few original settlers left alive. He was born April 15, 1915. At eighty-six (in 2001), he is spry and active. He still does a full share of farm work, handling 15-kg sacks of habanero chiles as if they were feathers. His mind and memory are clear and sharp.

He and his older relatives in the Xool family hailed from Chunhuas, which was then in a slightly different place from today; the community moved to be on the highway. He lived briefly in San Diego, later called Ignacio Najera, but virtually the entire population of this community moved to Chunhuhub.

José and his brothers and cousins started coming to Chunhuhub in 1933, as chicleros. They settled and farmed a bit, gradually turning the place into a modest forest settlement. Contact with the outside world was by horse and mule over rain-forest trails; they had to haul the chicle all the way to Peto. The first truck road was one they cut to Morelos, around 1938, more or less following the route of the cemetery road today. The first truck brought fireworks for a celebration!

Festivals were simple and traditional. *Ch'a' chaak* (ceremonies to pray for rain) were held at the entrances to the pueblo, while *loj* (other, lesser ceremonies) were held in the center. The music was a violin, a drum with a deer-hide head, and a small drum headed with agouti hide. Bullfighters used deer hides for capes. There was plenty to eat; they would kill a steer or five or six pigs, and provide maize foods and chicken stew in proportion—to say nothing of raw rum in huge jars.

In those days, deer, turkeys, and guans were there for the taking. Animals no longer present, including tapir, brocket, and white-lipped peccary, were common. There were many monkeys. The soil was extremely fertile; onions reached half a kilo, and a jicama (tuberous root of a native bean species) weighing 15 kilos is still remembered.

Soil fertility declined with farming, while pest populations skyrocketed. Today, many crops once grown, such as soya and sorghum, no longer flourish. Needs for fertilizer, pesticide, and irrigation render the less profitable crops uneconomic. Citrus has come, needing much less care, and there are now some 5,000 ha of citrus in the ejido.

José served as delegado (the equivalent of mayor, for a small town) for sixteen years, and as comisario ejidal for a three-year term. (Town government and ejido government are completely separate entities with different functions; the town government maintains civil order and represents the town to higher units, while the ejido government manages the land and its resources. Only the heads of old settler families are official members of the ejido, while all Mexican citizens can vote for and participate in the town government.) During these terms of office, José agitated for schools and had much to do with bringing teachers, schools, and eventually a technical higher-level school (CEBETA) here. In this, one of his close associates was Don José García Michel, ancestor of José García García, the owner of the sawmill and of Rancho el Corozo; Michel (as he was known) started as an ordinary chiclero and built up the ranch by hard work, eventually becoming the town's "rich man."

Many of the huge trees that once crowded central Chunhuhub were consumed in burning lime and making ash for construction. A few old *pich* (*Enterolobium cyclocarpum*) trees, 2 meters thick at breast height, were left as landmarks; they show what the forest once was. No one in those days thought of conserving trees and animals; interior Quintana Roo was a vast wilderness. Unnecessary killing of trees or animals was, however, condemned, so the forests and game held up well until the 1990s. Chicle was usually the major cash source. It brought 70 centavos a kilo in the old days.

For a long time, chicle gathering, logging, and subsistence farming were the only activities. The great actor and singer Pedro Infante is said to have passed through when he was a young, unknown, impoverished chicle gatherer.

By 1970, Chunhuhub was a fair-sized town, and the citrus industry was beginning. It now provides much of the cash flow in the area. Cutting of railroad ties was important in those days, as it was until the late 1990s in some other ejidos. Cutting of valuable tropical timber was also important, and remains so, though the best has now been creamed off.

Prosperity came with roads and logging. Chunhuhub's main industry, the García sawmill, was founded about 1981. Migrants from as far as Nicaragua and El Salvador came to farm Chunhuhub's fertile lands and crop its wealth of precious woods. In a modern town where computer courses are taught and where teachers and technicians are a major "export," only the massive ruined church and the tiny, forgotten Classic Maya mounds still remind us of the long millennia of history.

Chunhuhub grew at a disproportionately high rate, because of its fertile soils and location at a strategic point on the major road. Most of the migration is from the nearby eastern and southern parts of Yucatán state. Here population is dense and land is scarce and worked-out. Others have come from coastal Quintana Roo—some as refugees from hurricanes such as 1989's Gilbert. Others have come from farther away; there are families from as far afield as Michoacán and Puebla. The vast majority of the in-migrants are Maya, sharing a common culture and social world with the native Zona Maya people. The few non-Maya families have been reasonably well accepted, in spite of being outsiders.

Between 1991 and 1996, the Mexican economy suffered a meltdown. The average worker's wage, in Mexico, began to decline in 1982, following debt crisis. It continued to slide, and finally crashed in 1995; during that year it fell fully 20 percent (*Los Angeles Times*, January 11, 1998). As of 1980, the Mexican worker earned 22 percent of the wage of the United States worker. As of 1996, both had lost ground, but the Mexican worker had lost far more and was making only 8 percent as much as an average Stateside worker.

Chunhuhub felt the shock. Most people were buffered from the worst of it, because they were subsistence farmers or cultivators of high-value fruit for tourist markets; still, the economic decline was serious.

By 2001, a rally had occurred. Five years of uneven growth had made it a much richer place than it was in 1991 or 1996. It had become modernized in many ways. Telephone service, somewhat reliable mail service, frequent taxi and bus service, and better roads had opened it to the outside world. Chunhuhub is now a prosperous place by southern Mexican standards.

In 1993 the village became a real town, an *alcaldía*: it elected a mayor (*alcalde*) instead of a delegate (*delegado*). Chunhuhub is unusual for a town of its size in that it is not the seat of a municipio (the Mexican equivalent of a county). It is a far outlier in the huge municipio of Felipe Carrillo Puerto. This situation exists because of the recent and rapid growth of the town. The entire municipio, in 1950, had only 8,320 people, most of them in Felipe Carrillo Puerto itself or in villages very near it. Chunhuhub at that time had barely been settled. The municipio grew to 32,506 people by 1980, and skyrocketed to 47,234 in 1990; much of the growth was in Chunhuhub, which, by then, had 7.3 percent of the people in the municipio (INEGI 1994:15–17). It is possible that, in the future, Chunhuhub may be split off in its own municipio.

A sad major event in Chunhuhub's recorded history came in September 1995: Hurricane Roxane passed directly through the town. Twenty-four hours of violent wind beat the town mercilessly, felling the great trees in the public square, blowing over houses or unroofing them, and devastating vast tracts of forest. Then came the rain: two weeks of it. The land could not absorb it, and rivers roared through town and country. CEBETA was under water—even the roofs. The computers there, Chunhuhub's pride, were ruined. (They were replaced early in 1996; the town insisted on this being a top priority in rebuilding.) More serious was the destruction of half the maize crop and almost all the commercial citrus crop. Traditional maize varieties planted in the traditional way—on the hills—did fine. Hybrid varieties in rich valley soils were wiped out by excessive water.

Thus one of Mexico's richest ejidos was reduced, literally at one blow, to poverty. When I arrived in 1996, many once-successful families were facing real hunger. Some were living largely on root crops, always the Maya's backup against such disasters (see, e.g., Terán and Rasmussen 1994). There was no actual starvation, thanks more to the highly successful farming of the town than to outside assistance; even half a harvest was a lot of corn.

Political fights in 1993, drought in 1994, and hurricane damage in 1995 left Chunhuhub in 1996 a much soberer and less optimistic place than it had been in 1991. Recovery did not begin until mid-1996. Chunhuhub was protected from the worst of the depression by its rich subsistence agriculture and by the proximity of Cancun—one of the few economic bright spots in Mexico during those years. But even in Chunhuhub, rising prices of purchased necessities and falling prices of local products caused massive disruption and grief. Not until around 1999 did the town recover. By 2001, however, it was far richer than ever, and public confidence was correspondingly high.

Chunhuhub is now (2004) a town of more than 6,000 people. The census of 1990 (an undercount, but not a huge undercount) found 3,453, with 5.6 persons per household. I counted about 4,500 in 1996 and nearly 6,000 in 2001. Presidente Juarez, the other town in which I did considerable research, had 940 people in 2001 and has not grown much since.

In plan, Chunhuhub is typical of small towns in southern Mexico. A central plaza, flanked by the church, government buildings, and stores, opens toward a marketplace. Around or near the latter are medical facilities and the town's main industry, the García family's sawmill. Otherwise, the town consists of a large area of tree-shaded gardens, in which the houses appear only as scattered roofs; from the air, the community appears more forest than town. From the ground, however, one quickly sees that the great trees of the dooryard gardens shade a dense and fast-growing population. Children are everywhere.

Chunhuhub is prosperous as Mexican rural communities go, but there are problems. As of 1990 (INEGI 2000c:57), 17 percent of the population of the municipio reported no income, which means they were subsistence farmers. (As of 1980, about a third of the adults in neighboring Jose María Morelos municipio had "no income," i.e., were subsistence farmers [INEGI 1987].) Another 34 percent were making less than the minimum wage; most of these were subsistence farmers with rural part-time work added, but many people with limited abilities or education survived on almost nothing in spite of working full time at odd jobs. Another 21 percent received the minimum or up to twice that lowly figure. Only 1.8 percent got more than five times the minimum. Moreover, some 10.3 percent were "unspecified," and these were probably yet other people in the subsistence economy.

On October 27, President Vicente Fox announced that although Mexico had become the tenth largest economy in the world, 40 percent of the population live in poverty, half of those in extreme poverty (*Por Esto!* October 29, 2001, "El Estado" pp. 2–3, "'TLC plus' con Canada y EE.UU."). As one of Mexico's richest states, Quintana Roo is far less affected, but it has its poor, and many of them are rural Maya.

In the municipio, household size held steady at about 6 per household (usually, in this area, a compound), but shrank to 5.5 in 1995 (INEGI 2000c:17) and has stayed about there since. This indicates a slight alleviation of crowding, but also a slight decline in the old compound system, with large extended families occupying adjacent houses in one walled enclosure. The compound system continues without change in the marginal parts of Chunhuhub and in the small forest settlements like Presidente Juarez.

As of 2001, jobs, prosperity, and progress had come, via spillover from the tourist boom. Statewide, the minimum wage rose. In 1999, it was 29.70 pesos per day, and only 3 percent make less; 21 percent make about the minimum; the vast majority of workers make much more (and there are virtually no "unspecified"; INEGI 2000a:155). No tourists come to Chunhuhub, but money from both federal and state governments has poured in, providing jobs in teaching, road work, and many other services. This money is recycled through dozens of small shops, workplaces, and services. Almost any enterprising villager can find something to do beyond farm labor. Everyone has decent clothes, and at least half the families are well dressed and relatively stylish. Virtually everyone has television, most have radios or other electronics as well, and many have refrigerators, stoves, fans, and other modern conveniences.

The total population of Felipe Carrillo Puerto municipio is now more than 50,000; Chunhuhub is the second biggest town, but it remains a long second after the municipio seat, Felipe Carrillo Puerto. Bordering ejidos include Polyuc, Dzula, Lopez Mateos, Emiliano Zapata, and Laguna Kanab. Ranches and national lands lie on the southern borders.

The main highway is now excellently paved. Some side roads are surfaced with *sascab* (partially decomposed limestone). This surfacing erodes continually, to the great annoyance of the community. Therefore, a few roads have been paved—always just before elections! Most roads are still dirt ruts, in which pigs wallow after rains. Rough branch roads lead to hamlets in the interior; the major branch routes are graded and partially surfaced, thus earning the still-current classic name of *sakbej*, "white road."

Many of these roads lead to uninhabited ejidos. Many large parcels of fertile land in the hills behind Chunhuhub were granted as ejidos, but are (or until recently were) not provided with water. The ejidatarios and their families thus reside in Chunhuhub and commute by bicycle, truck, or taxi to their lands. Since Chunhuhub is itself well provided with facilities and is a hospitable place, these people are welcomed, not least because they are good customers for the town's businesses. Also, they are Maya and often related to Chunhuhub ejidatarios. The people of four small forest ejidos actually live in Chunhuhub. These are Emiliano Zapata, with 15 ejidatarios; San Cristobal, with about 22; San Miguel, about 25; and San Rafael, about 8.

Chunhuhub itself has 333 ejidatario families, providing an average of 40 ha of ejido land per family. However, most of the town's people are not ejidatarios, being recent migrants from elsewhere, and they must rent land, usually at a low fee.

The town economy has flourished, but still Chunhuhub's young people have often joined the exodus to cities. This is especially true of the educated young adults, most of whom leave. Claudio Xool Huh, ejido president in 1996, estimated that 40 percent of young adults leave the town; this seems a bit high. The flight of the most enterprising and well-educated drains the town of promise. The situation is far worse in the poorer towns of northern Quintana Roo and eastern Yucatán state, which have lost almost 100 percent of their young adults to Cancun (see, e.g., Re Cruz 1996).

Chunhuhub, like most large Maya towns, has spun off some daughter colonies. The most successful is Margaritas; it has about a hundred cows, many sheep, and a great deal of corn, beans, and chiles. Occupying a lush, well-watered area of southwest Quintana Roo, it has excellent soil. It was founded by Mario Jimenez in 1972. He brought fifteen others. There are more than seventy-five families now.

The typical individual of the Zona Maya engages in several activities, as time permits. For men, farming is dominant, but beekeeping, logging, gathering, hunting, local cash work, brief migrations to cities for higher-paid labor or to sell fruit, and working for local ranches all are common sidelines. Women are primarily homemakers but also sell fruit, make tamales and similar snacks for sale, sew clothing, provide minor services locally, cut hair, and otherwise earn small sums working in or around the home. Many Chunhuhub families are not primarily farmers; teaching is the next most common occupation, followed by shopkeeping. However, these families usually have some land—at least a dooryard garden, but often a sizable orchard or even a ranch. There is, therefore, no hard and fast separation between teacher and farmer, storekeeper and rancher, urban and rural.

Diversity in occupation is not a desperate adaptation to poverty and marginality. The most well-to-do families in the village are at least as diverse in their adaptations as the poor. There is certainly a sense of spreading the risk, of insurance, in all this, but much of the diversification is done simply because it makes the most efficient use of time. Farming is a seasonal matter; short spells of sun-to-sun work alternate with long periods of relative inactivity. It is during these latter that people travel to town, make clothes, or work on government projects. Similarly, teaching is usually confined to half-day sessions on about 150 days out of the year, leaving most of the teacher's time free. In an ejido with land to spare, it makes sense to invest this time in developing an orange orchard, a small shop, or both.

The old colonial church remains the center of town. Next to the church on the north is the small chapel that served the Catholic congregation until the main church was restored. South of the church is the ejido hall, the current center of the most important civic activities. Behind the church, near the García sawmill, is the new radiotelephone—Chunhuhub's link with the world. It was installed around 1994.

Across the square, on the southeast corner, sat in 1991 the tiny *delegación* office—the official headquarters of Chunhuhub pueblo as opposed to Chunhuhub ejido. With the elevation to town status, the old office became a communications office, and an abandoned school just east of it was rebuilt into smart new town offices. A mayor's office, meeting room, police station, and other facilities appeared. In 2002, still further enlargement had occurred: Chunhuhub boasted a fine new town hall, with two modern computers. The former school had become an all-purpose building, with cultural center and offices.

The new hall houses the civil register (in which births, marriages, and deaths are recorded) and serves as the center and rallying point for the forces of law and order. Normally these consist of a few pleasant, relaxed policemen, who stand guard there or patrol the square. There are also the ancient guns of the *guardia*. This institution, a survival of the Cruzob Maya organization, is a local militia group that theoretically meets to drill, in order to remain prepared to fight off hostile government forces. The guardia is moribund now, but not totally abandoned; one can never tell what the future will bring, and any attempt to take away land would be resisted.

Almost everyone in Chunhuhub is poor, in the sense that they have little money and few chances to get more. Almost no one, however, is destitute. The poorest are the few landless families that must live by day labor: weeding, harvesting, construction. They are the only ones who are commonly stressed by absolute lack of food and clothing.

The vast majority of the town, and of the surrounding ejidos, falls into the realm of ordinary small farmers. They have enough land and resources to eat well and live comfortable (if rather spartan) lives most of the time, but a catastrophe like Hurricane Roxane can bring real hunger. Felix Medina Tzuc and his family represent this segment of the population. They lived in a pole-and-thatch house until 1998, when they could finally construct a masonry house. They usually eat well but simply. They were reduced to real want by Hurricane Roxane; their crops were ruined, and they lived for months on sweet potatoes, which, being underground, survived the storm. Their clothing was often threadbare until the sons got jobs in the cities and Don Felix was elected justice of the peace, a position with an appreciable salary. The aftermath of Hurricane Roxane reduced at least 20 percent, perhaps 25 percent, of the families of Chunhuhub to very straitened circumstances. Felix Medina Tzuc and his family were not alone in living on weevily corn and sweet potatoes for several months after Roxane.

The percentage of town residents that is genuinely needy is about 10 percent, almost all young families without access to ejido land (my survey data). This contrasts with an "extreme poverty" level in Mexico in general of 20 percent (Mary Beth Sheridan, "Welfare Is on a Roll in Mexico," *Los Angeles Times*, January 7, 1999). The "poverty" level in Mexico subsumes 36 percent of the population, but it is reckoned on income, and thus is hard to apply realistically to Chunhuhub. A good subsistence living in the Quintana Roo forest is much more comfortable than what is suggested by the phrase "no income."

The poverty is hard to see in Chunhuhub, because, as in most Mexican towns, the well-do-to live in the center and along the main road. The poor live at the edge of town, often hidden from roads in little pole-and-thatch compounds that must be reached by forest trails.

The government's "Progresa" program is administered through local women's groups. Chunhuhub has a very active women's group. They manage government grant money that is intended to go to the poorest families for food and clothing—the latter being primarily aimed at providing decent school clothes for children. (Uniforms are the rule in Mexico, which saves families from the competitive fads that make back-to-school shopping so stressful for many United States families.) Grants of a few hundred pesos a month are provided to about five hundred people.

Many cynically say that most of the problems are due to husbands who are lazy and drink too much. They are right for all too many cases, but many families—mostly recent immigrants—are sober, hardworking, and still desperate. These often do not stay long; they go on to try their fortune somewhere else. So they remain almost invisible. At least subsistence agriculture continues to provide a good living.

However, Progresa, which provides relief money for food and clothing to women in families, had identified in 2001 about a thousand people in Chunhuhub as needing some assistance. Typically, according to Progresa personnel, those they help usually eat well but lack some nutrients, and have clothing but not enough to send their children well dressed to school.

Poverty was somewhat eased in Chunhuhub until 1999 by the subsidy on tortillas. These are not only the calorie source of the poor; they are a vitally important source of vitamins, calcium, and fiber. The tortilla subsidy was removed at the beginning of 1999, but the government was forced to backpedal and try to keep tortilla prices low (James Smith, "Mexico Clamps Down on Tortilla Prices," *Los Angeles Times*, January 7, 1999). Moreover, at this time the Free Trade Agreement came into effect, dropping the price of maize, as cheap United States corn flooded the market. (This move ruined small-scale maize farmers all over Mexico, but did not impact those of Chunhuhub very badly, since maize was not a major sale crop; also, local maize is distinctive and preferred by the Maya, so it commands a sale even in the face of cheaper but less choice grain.)

Given the longstanding scorn in which elite Mexicans hold the tortilla (Pilcher 1998), it is well to emphasize the fact that the tortilla is exceedingly valuable nutritionally, in very dramatic contrast to the white bread, sodas, noodles, crackers, cookies, and candy that compete with it for the poor family's pesos. One can live on tortillas, beans, and chile. One dies on *comida chatarra* (the Mexican translation of "junk food").

Above the level of farmers are those who own large orchards, small shops, and the like. Still better off are teachers, larger shop owners, and skilled workers. The few "rich" of the town include the director of CEBETA, a few medium-scale merchants who trade in fruit and honey, and one or two others. These families are rich only by comparison with the majority of townsfolk. The only family that is at all rich by ordinary Mexican urban standards is the García family, with their sawmill, ranch, and other enterprises; they would rank as no more than upper middle class in urban Mexico, and they live a modest lifestyle, taking part in the community and not separating themselves as *ricos*.

Prestige and power in Chunhuhub are not simple matters. The García family is the only genuinely rich one, but they do not occupy high political positions; they are busy with their family enterprises. In contrast, the Xool family has high prestige and usually some people in high office; the Xools are ejidatarios descended from Chunhuhub's first settlers. The ejido presidents in 1991 and 1999–2002 had Xool ancestry; the ejido president in 1996 was a Xool; the town mayor from 2002 to 2005 was a Xool. Yet the Xools are not rich; indeed, some are quite poor. Moreover, they and other settler families are so intermarried with so many other families that they cannot form a solid bloc of "old settlers," though tension between ejidatario and later-coming families does occur and can affect decisions on land use.

The separation of ejido government from town government creates a de facto balance of powers. The community's elective mayor balances the ejido's elective offices. In short, Chunhuhub is economically differentiated. The García family has many times the wealth of a poor laborer's family. But the real power and wealth of Quintana Roo is far away. The entire community of Chunhuhub—all its property, income, and businesses combined—would seem tiny and insignificant to the great hoteliers, drug lords, and businessmen of Cancun.

Robert Redfield (1941, 1950; Redfield and Villa Rojas 1934) argued that Maya communities were fairly egalitarian, since everyone was far from affluent. One must agree with Goldkind (1965, 1966) in his critique of Redfield for underestimating the importance of local differentiation in wealth in Maya communities (specifically, in Chan Kom, Redfield's community in Yucatán state). But one must also agree with Redfield that the real difference is between urban wealth and rural poverty. Even the "rich" of Chan Kom and Chunhuhub are very small players on the Mexican scene. Moreover, rough ideas of social justice govern interaction and prevent excesses among the Quintana Roo Maya (Hostettler 2002). In a country that is rumored to have more billionaires than any other except the United States and Japan, a country where political power is sharply concentrated in a few Mexico City households, the differentiation in Chunhuhub seems

Elsi Ramirez cooking; note the k'oben, *hearth made of three stones.*

relatively unimportant. The powers who affect Chunhuhub life are the powerful players on the state and national scenes, not the local well-to-do families. A full political ecology of Quintana Roo is still to be written. When it comes, it will have to focus on international, national, and state policies; their effects on the local countryfolk; and how those countryfolk respond.

Education is another channel to prestige and relative affluence. The prestige of educated professionals like Dr. Velasco (the highly respected town pharmacist, now retired and inactive) and the director of CEBETA provides them with very high status on the local scene.

These circles intersect. Not surprisingly, given the fact that the Garcías are by far the town's biggest employer, there is mutual accommodation and rapport between the ejido government and the García enterprises. However, this is not as incestuous as one might expect from similar situations elsewhere. The ejido government insists on high prices for their wood, gets them, and is not at all afraid to face down the García mill on the issue. For his part, the actual manager of the mill, José García García, is one of the most friendly and sociable men in the village, and prefers to settle matters by talking things out. Similarly, the prestige of CEBETA and its staff did not prevent it from being attacked by the entire ejido government when CEBETA personnel in 1991 allowed a small fire to get out of control and burn some ejido fruit trees. There was a serious threat of taking away much of CEBETA's land, which is, of course, leased to it by the ejido. Only apologies, border demarcation, promises of good behavior and a bit of financial remuneration saved CEBETA's marginal acreage. The politics of a student strike (over irresponsible behavior by a few CEBETA teachers) in 1996 were also revealing; several of the student leaders came from important families, and the silent but obvious backing they got from their parents and kin had much to do with the strike's success.

Democracy is the rule. The mayoral and ejido elections are hotly contested, and there are meetings, campaigns, and voting. Ejido assemblies encourage speaking freely; women as well as men take the floor and have their say, often at length and in full oratory style. (Public speaking is taught in the schools; the Maya may have needed no instruction—they have a historical reputation in this regard.)

Typically, a hundred to two hundred people appear for an assembly. About 80 percent to 90 percent of these are male, but women—usually with children—are vocally present. The full body of 333 ejidatarios is supposed to attend, but this figure is approached only at the annual meeting when land is allocated. When health is the issue, more women turn out and may make up close to a majority.

Many people prepare for the Sunday morning occasion by getting drunk. (Some of them get drunk for the occasion; others are just drunk anyway and enjoy the rest and show.) Rough and ready democracy is the rule. Leading families such as the Tuns and Xools do not present a united front; Xools, being numerous and diverse, are always evident on all sides of everything. The town mayor does not interfere with ejido politics.

By contrast, the municipio head and the state governor are essentially appointed by PRI (Partido Revolucionario Institucional, Mexico's longtime ruling party). Quintana Roo was almost a one-party state up until 1997, and central Quintana Roo still is. Until recently, selection of PRI candidates was by party *dedazo*, not by voting. *Dedazo* means "big finger" and is the Mexican term for PRI's appointment process, in which the appropriate standing committee of the party picks the candidate; in a one-party polity, such as Quintana Roo in the early 1990s, this candidate automatically wins. Today, there are primaries—but here too, until recently, the winner was certain of victory in the full open contest. These primaries were somewhat nominal; they pitted the party's choice against PRI upstarts; the party's choice had the funding and organized backing, and always won. This, however, has now changed. The upstarts are mounting more and more credible threats, and the final elections are being seriously contested by other parties.

The politics of violence and factionalism that dominates so much of political life in Chiapas and Oaxaca is unthinkable in such a peaceful town as Chunhuhub, where the value on friendliness and accommodation is high. Threats of violence by caciques are inconceivable; people are willing, even eager, to stand up for their rights; and the community runs smoothly. Power is diffused, and no one group has a monopoly.

This does not mean, however, that conflicts do not occur. It does not mean that there is no tension, or that everyone is uniformly honest and unbiased. Still, the ejido is powerful and tightly organized enough to deal on a very strong footing with municipio and state governments. Large, militant, backed by powerful local interests, and run by a sophisticated group who are not unaware of their Caste War heritage, the ejido is not at all like the small and weak ejidos that can be pushed around at will by larger entities. (Some such ejidos exist not far away and serve as cautionary examples to the people of Chunhuhub.)

Probably the worst conflict in Chunhuhub's history developed when Chunhuhub became an alcaldía. The first-ever mayoral election was hotly contested, and there were even fistfights between partisans—an unheard-of and shocking thing in the tranquil town. The former ejido president, Teodomiro Tun Xool (note the family lines), ran against a rather retiring young man from another well-known family. The rumor was that Don Teodomiro—a fiery reformer—was too disturbing to the municipio government, and they had worked to get the more pliable opponent to run.

Don Teodomiro lost by a hair, amid the inevitable charges from both sides of double voting and miscounting. However, the town immediately settled back into its usual calm. Don Teodomiro opened a pharmacy instead of carrying on the struggle. The new mayor was regarded, even by his political opponents, as an honest and decent man; he was universally liked. The only problem was that he was widely regarded as too weak a leader to do much for the community; he and the ejido president were unable to stop the illegal logging that wiped out the ejido's financial base, or the poaching that has decimated game needed for the subsistence of the poor. The ejido president of the time, Claudio Xool Huh, complained bitterly to me of the inaction of officials and police in regard to illegal logging and hunting and the small but infuriating marijuana trade. The following year the ejido presidency changed. Illegal logging was wound down, to be thoroughly controlled by 2000.

The state and municipio governments had learned to view Chunhuhub as a "troublesome" community. This is said to have hindered the process of getting aid, especially after Roxane.

Mayoral politics can also become contentious. The 1996 mayoral campaign included one person—an educated outsider—known for shady associations and shady dealings in liquor, illegal timber, and other commodities. He was pitted against two highly regarded local populist candidates of rural background. Unfortunately, the latter split the "good candidate" vote, and the dubious candidate won with a bare plurality. Chunhuhub thus continued on a course marked by lack of planning and enforcement. Yet, once again, term limits contributed. Terms are three years and reelection is not permitted. The next election was won by one of the most respected citizens in Chunhuhub, and illegal actions and disturbances of the peace were promptly reduced to a minimum.

As of 1991, the ejido was working for an orange juice plant, better roads, municipio or at least town status, telephone service, better electricity, a post office, more reliable water pumping, and better credit terms from banks. By 1996, the community had achieved *all* of these goals, except the juice plant, which had ceased to be a viable economic option. In 1996, stated goals were a return to being able to log precious woods and a (consequent) return to fiscal solvency for the ejido, as well as expanded educational opportunities. By 2001, these goals had been achieved.

The ejido plans and monitors timber sales, reserves, and reforestation; milpa cutting and burning; raising cattle on grass-invaded lands; and the rapid development of citrus along the main highway. Ejido officials have talked to the state governor and routinely deal with state and municipio leadership. Yet the ejido officers are not the rich and powerful of the village; they are ordinary farmers and small traders. They are not the weakest and poorest by any means, and one could easily cast them as "elite" in comparison with (for instance) many of the recent-immigrant families or the families from some of the small nearby ejidos. But on the other hand they are not in a class with the Garcías or the CEBETA director, or even with the Pat family, which owns the town's three biggest tiendas.

In the 1990s, actual ejidatarios did not form a clear group over against non-ejidatarios. Migrants from Yucatán state did not form an identifiable group. Migrants from other parts of Mexico were scattered about the village. Socializing was usually by extended family. Large kinship networks had built up—not only among the old settler families such as Tun, Xool, Cetz, or Canul, but even among recent immigrant ones. Relatives had often migrated together or followed each other, so that networks of kin had formed quickly.

The Mexican rural diaspora had also had its effect on the size of the social world. Chunhuhubians have traveled as far as Fort Bragg, California, and Kenosha, Wisconsin. Several Chunhuhubians worked or had worked in Los Angeles, and one had been tragically robbed and murdered there. A steady stream of Chunhuhubians was constantly flowing to and from Cancun, Chetumal, and Mérida (in that order of importance). Others were scattered over south Mexico. This gave more reality to the outside world visible on television.

By 2001, more people were leaving; more outsiders were coming in. Essentially every household had television. On September 11, we all watched the terrorist suicide bombings unfold before our eyes. The peaceful Maya reacted with horror and incomprehension.

Maya identity becomes ever more fluid, negotiated, and dialogic (Hervik 1999; see also Castaneda 1996; Re Cruz 1996). As people shift to speaking Spanish and marry outside the community, more and more individuals become *mestizo*. Then they can, and typically do, call themselves

"Maya" in the villages and "Mexican" in the cities. Anti-indigenous prejudice is weak in Quintana Roo, but visible enough to make some Maya wish to leave their heritage at the village gate. Others, however, only reassert it more strongly. Ethnicity becomes oddly structured. For one example, the ordinary foods of the Yucatán Peninsula were not specially singled out by restaurants in the 1980s; in the 1990s they were *comida típica* or *comida regional* ("typical" or "regional cuisine"); since the mid-1990s, they are "Maya" dishes. This change has been driven by the encroachment of foreign and Mexico City cuisines and the simultaneous valorization of "Mayaness" because of tourism and reasserted regional pride.

Family is all-important in Yucatec Maya life (Redfield and Villa Rojas 1934; Restall 1997). A person exists as part of a network of kin. The nuclear family is the focus of life in Chunhuhub. Most households are nuclear, and almost all the rest are three-generation families consisting of one or two elderly people living with a married child or children and several grandchildren. The very few other households are mere variations on the above: when the grandparents die, the household sometimes continues to house siblings and their various children for a while.

People say that nuclear family living is on the increase, but there are still many extended-family "compounds" of the sort widely reported for the Yucatec Maya. In Chunhuhub, a compound is defined both by its dry-laid stone walls and by official surveys. Extended-family compounds arose when a couple obtained a large amount of land, and could settle some or all of their children in their own houses on the land. These children have children of their own, and eventually a large three-generation or even four-generation compound results. However, when the old founding couple dies, the compound tends to break up.

Typical was a pattern in which essentially separate households lived in a compound, shared the trees, helped each other with work, and helped each other economically without pooling all resources. Another pattern consisted of adjacent compounds: parents and children, or brothers and sisters, living next door to each other. In these cases there is a great deal of mutual aid and support. Typically, the parents or older siblings arrange for the plots of land for everyone. Such arrangements are continually developing at the fringes of Chunhuhub. However, typically, siblings scatter over the village, wherever they find land. Moreover, since so many Chunhuhubians have migrated to or from other communities, there are wide-flung kinship networks. Grandparents come to visit, or are visited, for up to several weeks. Sisters return from Cancun, brothers return from Los Angeles. Children go to live with grandparents in Oxkutzcab for schooling. A man needing help calls in his brother from near Valladolid; next year he will return the favor when his brother needs a hand. A man wanting to plant citrus seeks out a cousin in Ticul—a center of citrus nursery activity—who will give him young trees below market price.

In short, kinship, usually direct lineal kinship, still structures life and overwhelmingly dominates in matters of economic support and mutual aid. This is seen in ejido politics. The original settler families have now grown large, intermarried for two generations, and come to know each other exceedingly well. That they dominate the ejido is expected. This being said, it is highly interesting that many successful farmers are not ejidatarios, nor are the highly educated and urbanized professionals. All these are relatively recent migrants. The original ejidatarios are overwhelmingly concentrated in traditional small-scale agriculture: milpas and small orchards. A few operate small shops. Many work for the sawmill or other wage-paying employers.

The kinship system is typical of Yucatec communities. It is broadly patrilateral and patrilocal, but female lines are important too. The kaaj, the community, is a group of patrilineal families, often related (Restall 1996). This is what Chunhuhub was before massive in-migration from other areas. The patrilateral bias but bilateral reality is shown by family names. These follow a pattern that apparently characterized both the Maya and the Spanish in the old days: father's and mother's names are both used, but the mother's surname is dropped after one generation. People almost always use both last names. Husband's name is attached by a *de* to wife's full maiden name; Aurora Dzib Xihum de Cen's father is a Dzib, her mother a Xihum, and her husband a Cen. Descent is more or less patrilateral, but with various accommodations (e.g., female-identified goods like linens pass from mother to daughter or daughter-in-law, and uncles on both sides of the family may leave or give items to nieces and nephews). The importance of the maternal family is notable. A person can count on full support from his mother's kin. Among households, mother's kin are as active in networking and interaction as father's kin. Thus, the networks of kin that pervade Chunhuhub and tie it to the rest of the Zona Maya are bilaterally reckoned. In the same spirit, sisters are important in family economic dealings. Extended-family compounds often include a sister with her husband, though brothers or unmarried sisters are more commonly involved. Some family compounds, in fact, are more matrilineal than patrilineal. Some, though patrilineal, reveal matrilocality; the Dzib sons and daughters have all held together, with daughters' husbands moving into or next to the Dzib family compound.

Extremely important is *compadrasco*, godparenthood. Neighbors and friends stand sponsor and provide financial help and willing hands at baptisms, weddings, and coming-of-age ceremonies—notably the all-important coming-out party at a girl's fifteenth birthday. Through this, they become *compadres* and *comadres*, godfathers and godmothers. Any close tie of friendship or co-work is bound to be formalized as compadrasco sooner or later. Since the need for cleaving to friends is great and the resources of the average friend are slender, a ceremony can draw on many friendships, and thus can result in many compadrasco relationships. One person supplies the soft drinks, thus becoming the *padrino de refrescos*. (*Padrino/a* is the term used by a godchild for a godparent, and is the usual term in these combinations—not *compadre*.) Another provides the cake, another the photography, and so on. Mexico has long been famous for its networks of "fictive kin." The custom has Mediterranean roots but also may have pre-Columbian analogues. Quintana Roo's Maya communities depend on such networks to maintain the wide-flung mutual aid that is necessary for survival in this risky environment.

Relative to what I am told about (and what I have observed in) Maya towns near Mérida, where the Spanish influence is more pronounced and the hierarchic society of the henequen estates is still alive, the Zona Maya of Quintana Roo appears markedly less patriarchal and more a stronghold of women's rights. In relation to most of Mexico, the entire Yucatán Peninsula is less overtly sexist. Men often make scornful remarks about the "machismo" of central Mexicans. This is so well known throughout Mexico that it is the source of countless jokes. (My colleague Michael

Kearney, whose field is Oaxaca, tells one: A big man from north Mexico says, "Back where I come from we are ALL REAL MACHOS!" The Yucateco answers, "Well, down here, half of the people are women, and we like it that way.") Certainly, in Chunhuhub, "macho" is definitely a pejorative term among both men and women. This is not to say that sexism and patriarchal behavior are unknown or even rare, but women certainly have a relatively high amount of public visibility and power compared to many places in Mexico or the United States.

Education in matters of agriculture and resource management is still informal, carried out within the family and work group. Parents are normally—and are expected to be—caring and gentle. They maintain household rules, but enforce the rules in a positive, friendly, relaxed way. No doubt related to this and to the pervasive down-valuing of violence is the lenient treatment of children. This contrasts with the situation in much of the old henequen zone of Yucatán, and immigrants often comment at length on it. "When I was young, the word was, 'If you don't castigate your children, Jesus will castigate you'"—so said one older lady, shaking her head at the way children are indulged in Chunhuhub and allowed to do almost as they please. (With her own children, although she administered a considerable amount of verbal discipline, she had acculturated to the local mores and allowed them to act more or less as they wished.) Some immigrants shared her negative opinion of permissive child rearing; others found it a relief and still resented the beatings of their youth. In any case, however, discipline was real and effective in Chunhuhub, being administered primarily by verbal scolding. Physical punishment was not common, though it was a well-known option. The parents' ideal child was one who helped parents and older siblings with household tasks, did excellently in school, and showed a restrained, quiet, grave demeanor in public. He or she respectfully obeyed parents and elder siblings, as well as teachers and other authority figures, and took sedulous care of younger siblings. There were many such children in town, and they came from families that appeared no more punitive than the others. Indeed, as so often happens in this world, many families included both "model" and "rebellious" children.

The superb work of Hilaria Maas Colli (1983), alas unpublished, is the one really brilliant light cast on Maya traditional education. It is a light focused on public ceremonies and rituals more than on everyday behavior. Fortunately, its insights and conclusions, based on a lifetime of exceptionally sensitive and thoughtful observation, lay a foundation for future studies. She has recounted in detail how a Maya girl gradually comes to be a fully participant Maya woman, aware of social roles and economic responsibilities. Among other things, she analyzes the *jets'mek'* ceremony, in which a baby becoming a toddler is first carried on the hip (theoretically at four months for a boy, three for a girl; actually any time during babyhood or toddlerhood). It is a very simple ceremony: parent or godparent carries the child around the room a few times. The child is then presented with toy farming tools if male, toy sewing and cooking tools if female. As Maas Colli points out, this sharply and clearly communicates gender-role expectations.

In learning about plants, animals, and agriculture, children follow their parents and elder siblings, and are instructed by them in the practices of ordinary life. They ask many questions, which are patiently answered, but much of the learning is by simple observation and imitation. However, systematic if informal instruction does take place, especially in specialized agricultural pursuits like beekeeping. There are still a few *jmeen*—traditional ritualists and healing specialists—being trained; this involves long apprenticeship and careful instruction, normally with one's father or uncle. This is partly out of an expectation that such things run in families, but, also, it is difficult to find the time and opportunity if one does not actually live with a jmeen. Moreover, there are those who suspect *jmeenoob* of witchcraft, and do not wish to learn their arts.

Children pick up animal and plant lore, as well as other local knowledge, while going along with adults to field and forest. They help garden, they help gather firewood, they help with the animals and plants of the dooryard garden, and they help with milpa cultivation. Parents expect them to help as soon as they are physically and intellectually able. They soon realize that the family's food and shelter depend on their doing so.

Parents and older siblings instruct children on what firewood to gather, what plants to pick, and so on. Much learning, especially of the "basics" (here not reading and math, but firewood and farming), is acquired via direct orders. Also, adults spend considerable time discussing the plants and animals in question, and the children listen, aware that this is important and that their status in the community depends, in part, on their knowledge of such matters.

Listening to stories is one way for children to learn. Hunting stories are an important part of Maya male discourse, and children listen intently to the lively, animated exchange of such stories that so often goes on in the street or over hearth fire or campfire (Llanes Pasos 1993). Logging tales, herbal lore, and agricultural knowledge flow along the same channels. No one tells the children to attend to these; they see that such information is important, they hear it exchanged in the form of lively and highly specific tales of local significance, and they have to act on their knowledge daily.

Perhaps especially important is that all the adults a child interacts with are apt to know such lore and tell such stories. Adults form groups, exchanging and sharing knowledge. Children take part to the extent that they play adult roles. Thus, initiation into the adult knowledge-sharing scene is an important part of progress to adulthood. People win status by knowing and by degree of incorporation into adult knowledge circles.

Learning is not a process that stops with childhood's end. Don José the healer still looks for new herbal knowledge. Don Felix constantly seeks out new forest lore. Today, many town children find that knowledge of machines, of schoolbook learning, and of business is more important than the old wisdom of the forests and fields. However, they learn mechanical and shopkeeping wisdom by the same methods, and bring to all their learning the same respect.

Levels of education are rising very rapidly in rural Quintana Roo, but isolated communities often lag behind. In remote settlements—and even in the rural fringe of Chunhuhub—many parents still consider a girl well enough educated if she knows the bare rudiments of reading and writing. Schools—especially in the isolated settlements—often meet for only three hours a day, and that on fewer than half the days of the year. Elementary schoolchildren typically have half-day sessions. Girls often stop school after two or three years, and in the remote settlements they rarely go beyond sixth grade. Boys are more apt to go on, but even boys often stop at third grade, and many stop after sixth. In Presidente Juarez, for example, this is the pattern. Some go through the village's high school (a very good one), but opportunities beyond that are limited, and few are positioned to take advantage of them. Many Presidente Juarez children are content to perform at a minimal level in school, sure that they will be small farmers or farmers' wives in the near future, needing only some rudimentary literacy and mathematics. Apathy, insolence, and even some drug use are not unknown in the schools. One teacher was rather amused at the children's attempts to hide their scurrilous talk by using Maya, since he, like most local teachers, is Maya himself and knew perfectly well what was being said.

Mexico's rural schools are well equipped and well set up. Adequate materials are produced on shoestring budgets; superb textbooks and learning materials are turned out by mimeograph on newsprint. The high school text *Dinámica social de Quintana Roo* (Ancona et al. 1995), for instance, includes a discussion of local ecology that would do credit to a major biological monograph.

Chunhuhub's patron is the cross of Jesus image in the old church, made of *chulul* wood. Now it is rather neglected; hence the town's problems—according to some very traditional believers. It predates the War of the Castes. It was guarded during that war, even when the settlement was deserted.

However, until recently, the people of the community were more concerned with the Yuntsiloob, the Lords. These are also called the *nojoch ts'uul*, the Powerful Ones, or the *nukuch p'ok*: the Big Hats. (The rather irreverent-sounding term "Big Hats" originally referred to the ceremonial headdresses of the ancient Maya gods; "great headdresses" might be a preferable translation.) Particularly important in a farm town like Chunhuhub are the Yumilk'aax, the Lords of the Forest, and the *yumilkool*, the Lords of the Milpa. People still sacrifice to them at many occasions, especially the loj ceremonies that are necessary for safety and prosperity. Jesus himself has fused with them; he is, in Maya, the *Ki'ichkelem Yuum*, the Handsome Lord—the old designation for the pre-Columbian maize god.

Supernatural beings in general are called *ik'*, "winds" (so pronounced in Chunhuhub, but more properly *'iik'* in Maya.) These include the *yuum*, the chaak (rain gods), and various other spirits and traditional entities. God

(*Dios*) and the Catholic saints are not usually included in the term. However, the line between Maya deities and Catholic saints is blurred by centuries of syncretism (cf. Redfield and Villa Rojas 1934; Villa Rojas 1945, 1978). The local guardian of deer is the god *Siip*, as in pre-Columbian times (he is the "*zip*" of Fray Diego de Landa's sixteenth-century description of Mayaland; Landa 1937). But the protective patron of peccaries is St. Sebastian, who was martyred by being shot with arrows (as peccaries were in the old times), and the protector of hunting and game in general is St. Martin the Hunter. Presumably, St. Sebastian and St. Martin stand here for pre-Columbian gods with Maya names.

Syncretism has not been a simple or mechanical process. Different parts of the peninsula have different syncretistic beliefs. The Cruzob Maya have their own creative fusion of traditions. New ideas arise daily in the contemporary world, as Maya bring old beliefs into new religious forms such as evangelical Protestantism.

Commonest of 'iik' are the *aluxes*, or *'aluuxo'ob* (the former, Hispanicized plural has become the common one). These are tiny elflike humanoid beings that guard field and forest. They are often visible as small Classic and postclassic Maya figurines of stone or pottery; these figurines hide in ruins and tree roots and come alive at night to do their business. Seeing an alux is usually safe, but if it passes to one side of a person, bad luck is indicated. Ariel Gongora saw an alux in Chunhuas—a tiny man half a meter tall, in a huge white hat. This was scary but a good sighting. He also saw a red flame near his house—not a fire but a *mal aire*, indicating treasure around there. If you watch such a fire steadily and pray, it will turn into an animal or bird and tell you to pray again, then show you where the treasure is. A man saw a calf in a cave in Oxkutzcab. The calf shrank away before his eyes. Where it had stood, the man found gold and silver—three pots. At Tepich is a cave with a treasure in it, but those who enter never return, making one wonder how anyone knows. Closer to home, at Chunhuhub's own Rancho Tampac, a bull's head emerged from a well, a mysterious calf was seen, and three pots of treasure were found. The finder had to have a Mass said for each pot; otherwise it would have disappeared.

Such treasure tales are known all over Mexico. George Foster, in a famous analysis, showed that they were a way of explaining wealth without blaming the wealthy ones for gouging their neighbors (Foster 1964; see also Foster 1965). In Chunhuhub, where wealth is easier to get, this trea-sure-hunting is less favorably viewed. It is too close to black magic.

Ordinary animals may have supernatural qualities. Snakes may turn into deer, and vice versa. Snakes whistle when hungry to lure birds. A whistle at night or under a rock in a ruin is probably a snake. (I believe it is the snake's hiss, echoed and "sharpened" in sound by the stone.) Some deadly snakes can roll along the ground with their tails in their mouths. These old beliefs still live, mainly among people over forty.

Until recently—and still today, in highly traditional towns like Xocen (Love and Peraza Castillo 1984; Terán and Rasmussen 1994)—Maya soci-ety was tied together by a constant round of rituals devoted to the Lords, the Catholic saints, and the various elves, spirits, and strange forces that inhabit the landscape. These rituals and their role in maintaining commu-nity have been described so well and so often that it would be tedious to rehearse them here (see Faust 1998; Love 1991; Love and Peraza Castillo 1984; Redfield and Villa Rojas 1934; Terán and Rasmussen 1994; Villa Rojas 1945). However, it is necessary to echo the point made in all these sources: the rituals tie the farmers to the land and to each other, and they regulate land management. They both create a management community and encode rules for managing. Some of them explicitly order conserva-tion, notably the hunting rituals (see Anderson and Tzuc 2004). Others, including the rituals for beehives and for clearing milpa land, are closely tied to concepts of wise use: taking only what one needs, and truly caring for it (in both senses of the phrase). Key in this is the degree to which rituals engage the emotions of the believers. Emotions are stirred, and are brought into harmony, so as to serve the social enterprise (Durkheim 1995 [1912]).

For the Maya, human society is intimately involved with plants and animals. The landscape is a human creation and is constantly being modi-fied. There is no opposition of "man" and "nature"; the world is a garden. Religion both expresses and constructs this (Anderson 1996a; Terán and Rasmussen 1994). Thus, it has been instrumental in maintaining good management and in keeping people together in their endeavors to manage for the long term.

This is expressed, to varying degrees, in the symbolism of particular ceremonies. Many of the most specifically management-oriented of these

Carnival in Chunhuhub.

have become otiose in Chunhuhub. However, traditional loj—based on pre-Columbian antecedents and devoted to the local Lords—still go on. Ch'a' chaak ceremonies still occur in some outlying settlements but have been dropped in Chunhuhub. (See full account of one such in the neighboring tiny settlement of Dos Aguadas, Anderson and Anderson 1996.)

For a *waajijkool* (one of the important loj, the "bread of the field"—which can be an orange orchard as well as a milpa), Don Rodolfo (universally known as "Don Duyo"), Chunhuhub's *jmeen*, officiates. The pib is dug, wood is burned over rocks till they are white with heat, and leaves are laid over them. Ritual cornbreads called *tutiwaj* (or *cho'okoj*), wrapped in leaves, are laid on this layer and all is buried. Smaller ones have five dots of sikil (squash seed meal), in a quincunx shape; larger ones have thirteen spots and a cross traced in sikil; still larger ones are layered, canonically with six layers of sikil between seven of corn dough to symbolize the thirteen levels of the Maya universe. Don Duyo has been in Campeche and knows something of the belief there that corn dough signifies flesh and sikil blood (Faust 1998), but he does not adhere to this symbolism. Meanwhile, someone makes *k'ool*: chicken stew thickened with corn dough and colored orange by achiote.

A simple *mesa* (field altar) is set up in the field or orchard. It is decorated with green leaves—usually of ja'abin, associated with water and life. An arch of ja'abin twigs protects it. It is graced with *santo 'uk'uj* (or *'uk'ij*; pozole sweetened with honey), and *baalche'* (honey mead flavored with bark of the *baalche'* tree). More often, today, honey is replaced by sugar, and the baalche' is little more than sugar water. The ritual breads are sprinkled with baalche' when they are unburied. Some are then arranged on the sides of the altar; the bread with the sikil cross lies in the center. A cooked chicken and five hard-boiled eggs are added. Most of the ritual breads are broken up into the k'ool.

Don Duyo lights a candle, then others from the first one. The adult men present each take a candle and go to the far side of the altar to pray.

Unburying ritual maize breads at a wajikool *("bread of the field"), a* loj *ritual for the welfare of a field or orchard. The breads are offerings to the* Yuntsiloob *that guard the field, the* yumilkool *(Lords of the Field).*

The women and children remain in the background. Don Duyo prays, rapidly chanting several *payalchi'oob* (Maya ritual orations; see Montemayor 1995; Terán and Rasmussen 1998). At pauses in these, baalche' is scattered across the altar, and Don Duyo makes crosses with a chicken's foot.

This completed, people eat the k'ool with its ritual breads, drink pozole and beer, and take some more home with them. The field is fed.

Even evangelical Protestants still hold these ceremonies, now glossed as thanksgiving celebrations (*actas de gracia*; the Spanish phrase has long been used to translate *loj*) dedicated to God. Even by Protestants, the old Yuntsiloob are mentioned and remembered. Traditional pib ovens are made, and pigs or even whole steers are cooked. The mesa is set up, with its corn gruel and chicken stew and its sacred flowers and leaves.

There are still those who divine with salt or corn. Twelve small piles of salt on an altar are left for a day or so; those that absorb much water indicate which months of the coming year will be wet.

For maize-kernel divination, 84 or 105 kernels are counted out. Don Marcos Puc Bacab uses an altar with a Christ sculpture, crucifix, and picture, and pictures of saints. He takes out his *saastun* (magic crystal), a clear glass marble, and lays it on the altar. Working with 84 kernels, he cuts the pile roughly into halves. Then he cuts off piles of 4 kernels until there are leftovers. This is done three times. If there are many leftover groups of one or two kernels, it is bad; one indicates that one is going alone too much, two indicates danger. Piles of three signal that faith in the Holy Trinity will prevail.

Other diviners, at least in the past, had more elaborate rituals.

Chunhuhub's annual fiesta is small; the Catholics carry the Christ through the streets, and there are firecrackers and some music. Other smaller and more isolated villages have preserved the Maya fiesta tradition in much more elaborate style. Bullfights, heavy drinking, and altars with wax and paper flowers are the rule. Maya music (*maayajpaax*) is played by older musicians. Typical are fiddle—sometimes played off the shoulder in medieval style—and drum; there is usually a saxophone or trumpet or other band instruments. Women advance with candles and pray, while men officiate and organize activities. Maya fiestas have been described in detail in many books, and western Quintana Roo has afforded me nothing to add to the descriptions therein. Don Felix, modern and practical, still reminisces about the days when Chunhuhub had such ceremonies. Of surviving ones in small towns nearby, he says: "They are doing this as their fathers did, as their grandfathers did, as their ancestors did. It's the custom." Anthropologists are not the only ones who meditatively romanticize a timeless past.

Even for those over forty, religious modernization has come. One result is that, today, Chunhuhub is divided between Catholics and Protestants. There has been no serious strife or intolerance over this issue. Catholics and Protestants mix happily; families often contain both. This situation contrasts sharply with the religious tension common in many parts of south Mexico. Strife between Catholics and Protestants has reached serious proportions in some areas. In highland Chiapas, Protestants have been expelled en masse from Chamula, Tenejapa, and other communities. Even in the immediate vicinity of Chunhuhub, such things happen: Xocen, a previously peaceful Yucatec Maya town not far away, expelled its Protestants. (As in Chiapas, this had much to do with preexisting tensions that had led one village faction to convert in the first place.) In Mérida, one leading peninsular newspaper—militantly right-wing Catholic—continually carries scathing articles about the *sectas* (non-Catholic denominations), which it accuses of everything from gross immorality to being entirely made up of U.S. secret agents and their dupes. To the editors of this newspaper, only one branch of Christianity is "religion"; all else is "sects." The same paper, at the height of Mexico's depression in the early 1990s, devoted a tremendous amount of space to Catholic denunciations of North American New Age spiritualism, while saying nothing about Mexican hard times.

This suggests some of the reasons why Maya are turning Protestant. In fact, the extremely rapid growth of Protestantism in Latin America is a complex and poorly understood phenomenon. Reasons given to me in Chunhuhub include the role the Catholic church played in the conquest, and the behavior of Bishop Landa subsequently. The Maya have long memories. More recent reality figures more importantly than Bishop Landa, however. The Catholic establishment in Yucatán has undeniably played a very conservative role: upholding paternalistic Hispanic elites, defending the power of *patrones* over workers, and quite openly and explicitly opposing Maya indigenous and economic movements. Most Maya express opinions at odds with these.

More strictly theological reasons for conversion to Protestantism include the complexity of Catholic ceremonies, the cult of saints "instead of the One God," and the priestly hierarchy as opposed to "my talking directly to the Lord." A somewhat different theological problem has occurred in many situations in which recently resurgent Catholic puritanism has led to direct attacks on Maya traditional syncretic religion. Xpichil, a community near Chunhuhub, suddenly found itself confronted by a new circuit-riding priest who forbade ch'a'chaak and loj. They, in turn, forbade him to return. Since then they have had no formal Catholic minister, and they use the town church for ch'a'chaak and loj.

Conversely, Catholic loyalists appeal to tradition and the traditional way of life. They also appeal to the proven success of Catholicism, which they tend to see as having protected them and their world for a long time.

A more general issue is birth control. The Catholic church of Yucatán is militant on this issue, and virtually no Catholic body of literature observed in the peninsula fails to stress the evils of all birth control (even rhythm and abstinence are barely tolerated). This position is leading to a rather rapid flow from Catholicism to Protestantism all over Latin America, not least in Quintana Roo, where the demographic transition is well under way and is deemed necessary and desirable by an expanding segment of the general population.

A very large number of people join the more "temperance"-oriented Protestant sects explicitly to stop drinking and smoking. "And," says Mario Jiménez of Margaritas, "a lot of them don't succeed." Yet, a lot do.

The driving, intense, youth-accessible music of the Protestants, especially the evangelical denominations, is sometimes mentioned. Associated with this is the active youth outreach of these groups. The Catholics of Chunhuhub rose to meet this challenge from 1996 onward, providing their own amplified musical services and youth-group activities.

The most direct, personal, persuasive, and widespread appeals by both sides are to personal experience, especially the healing of family members. Anyone who has had a family member recover from life-threatening illness or accident is loyal to any religious domination credited with helping in such an hour of need.

The outsider with Durkheimian training may point to a wider social fact. (Emile Durkheim argued, famously and correctly, that religion is normally more an expression and mirror of social life than a revealed Truth or an individual peak experience; Durkheim 1995 [1912].) The old harmony and unity of the village communities no longer exists. When everyone was a farmer, and everyone relied on everyone else's help, it made a great deal of sense to tie together the community by tight bonds of religion. The world is different now. Occupational specialization, flight of young people to cities, and the tensions of modernization have led to rapid differentiation and fragmentation. This is not just an anthropologist's interpretation. Xocen used startlingly Durkheimian arguments to explain the community's loyalty to syncretic Maya Catholicism and to justify the expulsion of Protestant communicants (personal research, but see also Terán and Rasmussen 1994). Xpichil also used Durkheim-like arguments, as well as appeals to militant traditions (which include strong Cruzob ties), to stop Catholic priests from visiting the town to hold Mass. Conversely, Chunhuhubians make theories about modernity and diversity to explain, and sometimes to justify, following new denominations.

Yet, even when Durkheimian solidarity still exists and is quite self-consciously addressed as a purpose, justification, and source of religion, Protestantism is winning. The Protestants expelled from Xocen have formed two small communities near Chunhuhub, thus re-creating Durkheimian solidarity in a new site. More striking is the case of Presidente Juarez. Presidente Juarez is a traditional Maya town: a close-knit community of farmers. It is united by participation in the Church of God of the Prophecy, an evangelical denomination based in Tennessee but characterized by very large and independent Latin American branches. (It is also the largest Protestant denomination in Chunhuhub. Though evangelical, it has become a sober, quiet, formal, politically progressive church in Mayaland.) Presidente Juarez is made up of families that moved from Chichimila in eastern Yucatán in the middle 1950s. They were looking for an isolated place to practice their faith without temptation. Their Protestantism is uncompromising; one revealed his own values by asking me if it is true that some churches allow "vices like drinking, smoking, robbery, and murder."

The true home of Protestantism in Quintana Roo is not the villages nor the established urban areas; it is the shantytowns and migrant suburbs around the newly exploding cities. Here, Maya displaced from their villages find themselves among strangers, confronted by new problems and new ways of life. Far-flung networks of kinship and occupation replace the tight village bonds. It is in these new, impoverished, unsettled communities that every block has its small evangelical temple, Baptist ministry, or Jehovah's Witness hall. These create warmth, human contact, and support in what would otherwise be a wasteland (Higuera Bonfil 1997; Vallarta Velez 1997.)

This being said, sectarianism is not problematic in Chunhuhub. There are about two hundred active Catholics, perhaps three hundred evangelical Protestants, and members of several other small Protestant congregations. Most denominations have their own small churches. Thus, in addition to the Catholic church, Chunhuhub possesses two different evangelical churches, two different Presbyterian churches (one evangelical, one not; both very small), a Jehovah's Witnesses meeting hall, and a Seventh-Day Adventist building. Baptists briefly had a small church, but it closed. The few Mormons so far lack a formal meeting place.

If there is any tendency for one type of person to concentrate in one denomination, I was unable to observe it. Every sizable congregation runs the gamut from elderly ladies in *rebozos* (old-fashioned shawls) to teenagers and children in modern town clothes, and from subsistence farmers to prosperous, educated professionals. Each congregation seems to be a random sample of the Chunhuhub social scene. The predictor of membership is not social position but kinship. Converts bring in their families and relatives, and then, if possible, their neighbors and work associates. Congregations turn out to be linked by blood and affinal relationship into single groups or a few closely interlocked groups. The range of kinship solidarity runs from the town's Seventh-Day Adventists, who are basically one extended family, to the Catholics, who are the most diverse group in terms of family and network membership.

This leaves the other five-thousand-odd Chunhuhubians as only mildly religious, if at all. Most are nominal Catholics who never go to church. Many are nominal Protestants or are simply undecided. (In Felipe Carrillo Puerto there are a few people who say they are strictly following ancient Maya religion—worshipping the chaak rain gods and the other ancient deities. This is viewed with interest by some Chunhuhubians.) Some people are evangelicals one year and Presbyterians the next, or are not sure which sort of evangelical they are.

At Easter 1996, attendance in local churches reaches some 200 at the Catholic church; another 120 at the Temple of God, the main evangelical meeting hall; well over 100 at the Church of God, the other evangelical church; about 50 at each Presbyterian church; around 10 for the Jehovah's Witnesses (who do not recognize Easter as special); and only 5 to 10 at the Baptist site. A regular Sunday brings out about half this turnout.

Friendship, and indeed even compadrasco, does not stop at religious borders. Some people are members of two faiths. Protestants go to Catholic ceremonies, and vice versa.

Maya syncretism, always obvious in Maya Catholicism, is moving into the Protestant world. In 1991, the biggest *janlikool*—a quintessentially Maya loj—was thrown by the leading Church of God family, who had reinterpreted it to be a "thanksgiving to God" rather than a feeding of the milpa-spirit. About 150 people, from all religious categories in the town, attended. Singing, chanting, Bible reading, prayer, and eating took place in the field. Food baked in pib, or stewed, was set out on tables as offering. The usual pib breads dominated, especially the thirteen-layered piim that has clear and well-known traditional ritual significance. With it were served equally ritualistic thick atole with achiote, chile, and salt, as well as *sak k'ool* ("white stew," chicken cooked in white corn gruel) and *pollo* pibil (chickens basted with achiote and baked in the earth oven) as well as fruit, *makal* (yam root), atole, and sugarcane. People carried off buckets of food.

By 1996, Protestant groups were holding many such "acts of grace." These were traditional loj at which Christian preaching and electrified music groups replaced the chanting and altar ritual of the jmeen.

The old ceremonies are slow to perish. Ceremonies were always the property of experts. By 1991 they were already much reduced in number and elaborateness from those described in the classic ethnographies. Yet, in 2001, they survived—rare and unobtrusive, to be sure. My landlord is normally a very self-consciously modern and secular man, but I came home one day to find my room sprinkled thickly with a sour-smelling whitish substance. It turned out that my host had hired the jmeen Don Duyo to do a general blessing of the whole household, including my share of it, with *payalchi'*—traditional ritual *oraciones* (orations or prayers)—and sacred corn gruel. Don Duyo was there, his hands holding the *siipche'* leaves that carry blessing. And Don Marcos is still called away for such purposes as blessing a ranch that was troubled by Juan T'uul—"John the Rabbit"—the Maya trickster-imp. (He is a fusion of Juan—the human trickster of Spanish folktales—and the trickster rabbit of pre-Columbian lore.)

The breakdown of traditional religion should mean the decline of community, especially of the moral community. To some extent this has indeed happened. Not only have the old morals weakened, but they have been replaced by new ones, since a vacuum in this area would be hard to sustain. Modernization has brought not only new means of wreaking havoc on the environment, but also typical rural Mexican values, which are less environmentally conscious than Maya ones. Mexican values have changed greatly in recent years, and educated urban Mexicans are now at least as environmentally conscious as are similar groups elsewhere in the world, but this is not the side of Mexico that the Maya usually see.

Thus, environmentally friendly practices are less universal than they were in 1991, and community is in some ways weaker. However, old ways of farming go on. Hunting is still subject to rigid moral controls. This is partly because the old religion still has much force, especially among the older and highly traditional Maya who dominate milpa farming. The modernized Maya, who have lost the old religion, have become teachers and shopkeepers and taxi drivers.

Cooperation, mutual aid, and mutual responsibility have fared better than conservation. They seem not to have declined enough to be matters of greatly increased concern, though interviewees mention problems. Families teach these values; teaching does not depend on the old system of community rituals.

To some extent, Lloyd Warner's "civil religion" (Warner 1953)—patriotism and nationalist festivals substituting for, or used like, religious ones—has been successfully substituted for older rituals. Community solidarity values are openly celebrated and sustained through Independence Day (September 16) celebrations, school graduation ceremonies, and other civic affairs. Civil government and the ejido government have taken on new roles, new power, and new visibility. Civic culture may have less moral force, but it has more physical force. People can be jailed or fined for bad behavior. On the other hand, they are much less likely to be judged harshly for killing too many animals or for burning sapote trees. Civic ills are political ills: fighting, stealing, selling drugs. Civic goods still include mutual aid and tranquility, but ecology is no longer central among concerns. In regard to conservation, only the protection (by sustainable cutting) of precious woods has any force in the civic culture of today.

For thousands of years, the Maya have faced change. They have had their successes and failures, their rise and decline, but they have persisted. They have retained a cultural core—language, agricultural technology, religion—that still survives, adapting flexibly but never dying out. They are aware that they face some of the most severe challenges in their history. With cheerful pragmatism, but also with a dogged determination that draws on a painfully acute consciousness of centuries of hard times, they are meeting those challenges—so far.

The *jmeenoob* still have apprentices, and there are still extremely knowledgeable and wise farmers training sons in the profession, but it is clear that genuine experts like Felix Medina Tzuc or Antonio Azueta Chel will be rare indeed in the rising generation. Farming has changed; one needs to know pesticides and forage grasses and government policies; one does not need to know, and has little time to observe, such aspects of Don Felix's lore as the nesting habits of the Piratic Flycatcher and the method of calling the shy Pheasant Cuckoo until it is flying around one's head.

The romantics will find this tragic. The Maya too have their romantic streak, especially when it comes to the past, and they grieve over the losses. But they are forced into realism about the subject, and they are nothing if not good adapters. They created one of the greatest civilizations of the elder world, survived its fall, experienced Spanish conquest, maintained independence and culture in spite of it, and finally adjusted to the twentieth century. They did not do this by remaining nostalgically wedded to every vestige of the past.

Yet, they did not do it by giving up core values, either. What survives, besides the language, is a high value on tranquility and on a gracious, ritualized sense of life. Also still evident is the deep, quiet, but sincere religiosity, now coming out more through Protestant community involvement than through any traditional form. Most of all, the relationship with the land endures. Maya culture will survive for at least a few more generations.

Yet, playing against all hopeful currents of change is the older world of impoverished Mexico—a world perhaps created from tradition and isolation, as Redfield held, but maintained in actuality by the structural forces of oppression, which keep so many in poverty and powerlessness.

This is the world in which people say: *Hay que aguantar*, "you simply have to bear it."

> *Madre mia, cuando muera*
> *entiérrame en tu hogar*
> *y al hacer las tortillas*
> *podrás por mí llorar,*
> *pues si alguien te pregunta:*
> *"Señora, por que lloras?"*
> *dirás: La leña verde*
> *hace humo, hace llorar.*

> *(Bury me beneath your fire,*
> *Mother dear, if I should die;*
> *then if, as you make tortillas,*
> *you should think of me and cry,*
> *and if anyone should ask you*
> *"Lady, you are crying—why?"*
> *you can say: It's this green wood,*
> *it's the smoke that makes me cry.* [Zaid 1995])[1]

This children's folk song from central Mexico sums up a great deal of the experience of Mexico's rural and indigenous people. Life is hard, and until recently was usually short. Children could expect to die before reaching adulthood. People bore it stoically, showing as little emotion as they could. Giving way to emotion was too much to stand; it would lead to collapse. Countless millions of mothers fought back their tears.

The word was *aguantar*: bear, endure. Women, in particular, must endure hard and monotonous work, sickness and death in the family, and, all too often, drunken and abusive men, with little hope of any way out. Life went on, often getting worse but almost never getting any better, and above all never getting very different. Even now, in the small villages of

Quintana Roo, life is a daily round of making tortillas, feeding the animals, caring for the hordes of children—a round that has not changed much in five thousand years.

Today, the progress of health is such that people almost never have to deal with the death of children. First World people looking at the Third World never seem to understand just how horrible it is to lose a child. Many First World observers are lost in romantic fantasies of the good old days of traditional culture and simple life. Others, convinced (perhaps all too willingly) by the "smoke" excuse (in one form or another), believe that traditional people don't care, don't feel, and are used to loss. Typical are the words of the curate José Baeza, writing in 1813 of the Maya of Yucatán state: "It can be said of them that they are *sine affectione*, as Saint Paul said of the Gentiles . . . very rarely have I seen tears shed over the death of parents, children, spouses, or relatives. Much more commonly I have seen them with their eyes dry and serene in all circumstances" (cited and translated by Terry Rugeley 2001:20). Actually, of course, the sufferers were simply hiding their emotions—stoically bearing their fate, and certainly not admitting anything to the hated and feared Spanish. Nothing is more devastating than losing a child, and previous experience with such loss makes it worse, not better. Many of my older friends told me their stories; their heart-wounds were raw and agonizing decades after the fact. The myth of the insensitive mother is a racist fantasy. Yet one encounters this myth all over the world, even in some recent ethnographies.

The world of aguantar is still alive. People have to deal with what they have, and worrying about it accomplishes nothing except making it hurt more. There is much that is deeply noble and extremely courageous in such stoical endurance. It is the strongest and most difficult response of the human organism, though it so often passes unnoticed while weaker and noisier emotions attract all the attention. The cool, matter-of-fact way the child faces death in the song is indeed the way that Mexican rural and indigenous people face hardship. This should not be confused with lack of emotion. The whole drama of the song lies in the barely suppressed, or not-quite-suppressed, personal devastation that the mother and child feel. That too is a bitterly familiar part of life. Song allows people to transcend it to and wring some delight in creativity from the bitterness, but song does not solve the root problem of oppression and economic injustice.

The problem is that *aguantando*, carried too far, prevents doing what could actually be done. Rural and impoverished Mexicans are so used to putting up with the inevitable that they put up with the evitable as well. Health care could be improved. Garbage is scattered everywhere; municipal garbage dumps have finally come to Chunhuhub and many other towns, but they do not stop people from throwing their garbage around their yards or in the streets. Water is polluted and unsafe, yet few pressure the government for cleanup. Yards and gardens are often barely maintained. Game animals continue to decline from overhunting, though everyone understands what is happening; everyone knows it could easily be stopped if people would get together and agree on limits, but no one has much hope that that could happen. No one is happy with the results of all this, but no one has enough hope or optimism to do much about it. People endure.

Mexico is full of failed visions. The Maya states and Nahuatl empires fell. The visionary estates of the Franciscans, the reforms of the Bourbon monarchs, the anarchist and populist ideologies of the Revolution, and the modern plans and hopes of developers and political activists succeeded one another in the light, and followed one another into eclipse. Mexico remains a country where people work terribly hard, dream, have great visions of the future, try revolutionary plans. The people survive. The ultimate coping strategy for dealing with a dismal and unimprovable world is simply to bear it. But that strategy guarantees that the world will remain dismal and unimproved.

CHAPTER SEVEN

Medicine in West-Central Quintana Roo

MEDICINE in Chunhuhub today is closely tied to other aspects of human ecology, because ordinary medical care is still rooted in traditional Maya concepts of heating and cooling, and ordinary therapy is still herbal. Hundreds of plant species are drawn on for herbal cures. However, Mexico's comprehensive and effective public health and clinic system is active in the area and has convinced virtually everyone of its value. The result is a fascinating accommodation of Maya traditions and international biomedicine, with some Spanish magic added.

Maya medicine is conceptually very different from other medical traditions in Mexico, both international and folk. Moreover, it is not a single tradition; it is, indeed, a universe, in which people find various ways toward health.

The outside world has been impinging for a long time. Pre-Columbian influences from central Mexico and elsewhere must have been important, though we know little about them. In the sixteenth century, Spanish medicine (itself based largely on Arab developments of Greek medicine) was brought to the Maya, primarily by missionary friars. These, in turn, often proved surprisingly open to learning from the Maya (as the early dictionaries' entries on medicinal plants show; see, e.g., Arzápalo 1995). Ordinary Spanish settlers introduced many Old World magical beliefs. The resulting fusion produced its own complexity. In the nineteenth century, spiritualism became popular, but international scientific biomedicine also entered the picture, and it continues to gain ground today.

It seems reasonable to begin with a general account of modern Yucatec Maya medicine, then turn to matters of local belief and practice. I begin with emic views and end with an outsider's view—not precisely an "etic" one, since I am not so much applying a neutral etic grid as summarizing my own observations of the universe under investigation. (As anthropologists know but sometimes forget, "etic" does not mean "outsider's view." It refers to the application to a body of local data of a universal analytic standard, like a yardstick or a thermometer or the formal phonetic analysis that provided the origin for the word [phon-etic].)

In spite of all the problems listed above, the modern Maya of west-central Quintana Roo are currently a healthy population. Chunhuhub's cemetery remains tiny, and most of the graves in it display very long life-spans on their headstones. Only a few small graves indicate lives cut short before being fairly begun.

This is a new situation. Maya traditional medicine is a truly monumental intellectual and scientific achievement; even so, within living memory of local elders, half of all infants died and the overall life expectancy was around thirty-five. Alfonso Villa Rojas, traveling through the Zona Maya in the 1930s, recorded such figures, and noted the devastating effects of malaria and other stubborn diseases (Villa Rojas 1978). The elders remember those times and amply confirm what he said. They remember children, brothers, and sisters dying. Contrary to what many elite urbanites assume, the death of a child is not made any easier by the fact that lots of other children are dying (see chapter 6). The emotional scars of those infant deaths are raw and bloody and agonizing sixty years later.

What emerges from elders' accounts is a world not only of sudden death—with all its consequent emotional anguish—but of constant, chronic, crippling disease that held everyone down economically. From malaria and tuberculosis down to wounds and sores that never healed properly, the scourges of rural Quintana Roo were not rapid killers. They caused human misery far beyond the pain implied by the horrific mortality figures. Cutaneous leishmaniasis, transmitted by the bite of the chiclero fly, was endemic, disfiguring, and seriously incapacitating.

Today, life expectancy is high. The causes of death are those of the modern industrial world. Accidents, cancer, and heart disease are, in order, the main killers in Quintana Roo (INEGI 2000a:32). Diabetes is sixth; its incidence tracks the trend from maize and beans to sugar and white flour, thus measuring affluence and urbanization. Quintana Roo's infant mortality rate fluctuated from ten to twenty per thousand during the 1990s (INEGI 2000c:11), not much above the United States' rate. Mexico's as a whole was twenty-four in 2001 (UNICEF 2002), so Quintana Roo was doing very well indeed. Mexico's life expectancy in 2001 was seventy-three years, about the same as that in the United States; Quintana Roo's is unrecorded, but must be higher, given the far lower infant mortality rate. (Infant mortality affects disproportionately the average age of death).

For coping with what does go wrong, the people of the area have several options.

The first and by far the most important of these is the home. The vast majority of people self-medicate for the vast majority of conditions. Every adult (and adulthood begins early in the Zona Maya) knows the simple but effective dooryard cures. Spanish medicine gave mint, rue, oregano, and orange leaf teas for stomachache. The older Maya traditions recommend chaya (*Cnidoscolus chayamansa*, a green vegetable similar to spinach, but better tasting) for nutrition and as a diuretic, *k'anaan* (*Hamelia patens*) as an antiseptic wash, a drop of spiderwort (*Commelina* spp.) sap in the eye for minor eye diseases, *'elemuuy* root (*Malmea depressa, Annonaceae*) tea for kidney disorders (ranging from lower back pain to infrequent urination), and a host of others. Almost everyone knows that they should maintain some standards of cleanliness, such as boiling water, washing with strong soap, and bathing with antiseptic herbs, if there is any illness.

If home treatment fails, people go to an urban-trained medical doctor; if poor, they use the government clinics, but if they have any money, they usually prefer one of the private doctors. If this fails, or if they get referred, they go to hospitals in Felipe Carrillo Puerto or Chetumal; local clinics and doctors are not set up to treat serious diseases or chronic conditions or to do major surgery. Chunhuhub-area afflictions that are not trivial seem to me to be often very serious; referrals to the urban hospitals are relatively common.

One main concern of village doctors and midwives is assisting at childbirth—a frequent event indeed in these villages, where, until recently, almost every woman was either pregnant or nursing during most or all of her reproductive lifetime. Complications of birth are rare, and *partos* are normally routine. Local *parteras*—traditional birth attendants—oversee them (Anderson et al. 2004). Only when complications are expected will a woman go to the hospital to give birth.

The ordinary round of minor diseases, and the occasional major one, keeps the doctors fairly busy. The medical resources of Chunhuhub include, first and foremost, the government clinic. Every small town in Mexico has its *clínica*, staffed by a doctor who is doing a compulsory year of rural service after getting his or her degree. This is required as a way of paying back society for the costs of providing medical education. It is also an extremely effective way of getting medical care to the rural settlements. Clinics are spartan, but they have what really matters: they provide all the standard inoculations; they have antibiotics and other medicines; and they provide training and advice in sanitation, public health, malaria prevention, contraception, and other down-to-earth matters. The effect of such minimal care is truly incredible. It is, basically, this clinic system that has dropped the infant mortality rate by 98 percent in two generations and doubled the life expectancy. The benefits of this in terms of human well-being are not all captured by the statistics. The financial gain from eliminating loss of work time is in the millions. The psychological effects can only be imagined.

Private doctors occasionally settle briefly in the town, providing alternative care; they usually move on soon, but are replaced by newcomers.[1]

More salient, to many older people, are the traditional healers. Curers who also act as religious officiants (as mentioned earlier) are called *jmeen*, "doers." In this part of the peninsula they are not true shamans (cf. Eliade 1964; Faust 1998), priests, or healers, but they partake of the characteristics of all three; they are not well described by any non-Maya word. Chunhuhub's main jmeen and *yerbatero* (or *yerbero*, "herbalist") is Don Rodolfo (Don Duyo), but he is more jmeen than healer, and is old and only somewhat active. He sells herbs and patent medicines and holds minor ceremonies; in his younger days he was more active, and—among other things—served as a government-sponsored malaria control worker. He claims at various times to be 99 or 110 years old, but is actually about 70. Far more herbally sophisticated is José Cauich Canul of Polyuc, whose knowledge is revealed at length in another work (Anderson 2003). I worked many days, also, with

Marcos Puc Bacab of Presidente Juarez, a man of 66 (in 2001) who is an herbal and ritual expert and highly traditional; he speaks only Maya.

An old man on the trail said to me: "Every plant in the forest is medicine." To him, doctors know only "minerals dug out of the ground." The old Maya know the plants.

Maya medical lore is vast and sophisticated (García, Sierra, and Balám 1996; Anderson 2003). Dictionaries of the Spanish Colonial period, most notably the *Calepino de Motul* from around 1610, mentioned many herbal drugs (Arzápalo 1995). Various versions and editions of a mysterious work called *The Book of the Jew* (*El libro de Judío*) began circulating in the eighteenth century (Andrews Heath de Zapata 1979; Barrera and Barrera Vásquez 1983); whoever the "Jew" was (if he existed at all), he was certainly a convert to Maya medicine, for most of the book consists of directions for using well-known Maya medicinal plants. In the twentieth century, many ethnobotanists studied Maya herbal medicine. First was the pathbreaking ethnohistorian Ralph Roys (1976), who relied largely on the botanical identifications of Central American pioneer botanist Paul Standley. Later, the great Maya scholar Alfredo Barrera Vásquez collaborated with his botanist son Alfredo Barrera Marín and with Rosa María Franco in a major work, *Nomenclatura etnobotánica Maya* (1976). This was followed by a number of general ethnobiological works, and by a list of medicinal plants of Quintana Roo (Pulido Salas and Serralta Peraza 1993). Research has also been carried out in Belize (Arvigo and Balick 1993). Finally, Anita Ankli carried out major research in Yucatán state in the 1990s (Ankli 2000), working with her professor Michael Heinrich and with Otto Sticher to analyze and test Maya plant drugs in the laboratory (Ankli, Sticher, and Heinrich 1999a, 1999b). Comparable and really superb work on the Itzaj Maya, who are very close linguistic and cultural relatives of the Yucatec, is reported in the writings of Scott Atran and his associates (see Medin and Atran 1999 and references therein, especially Atran's article; some of his group also work with the Yucatec; see Atran et al. 2001). All this research is more or less related to the well-known work of the Berlins and their group in Chiapas (Berlin and Berlin 1996 and references therein).

I have recorded, so far, 347 named taxa that can be identified to species (or in some cases only to genus) level (Anderson 2003). Another 8 taxa can be identified to the level of family or larger group. Another 152 names have been recorded but have not yet been pinned down to species. This is less serious than it sounds, since all these names were recorded from two senior healers who tend to use names in an unconventional manner. Often, they used an odd name for a plant that was not in flower (or, occasionally, for one for which the specimen was lost). In such cases, the name could not be pinned to the species. I believe the total Chunhuhub-area pharmacopoeia is approximately 500 species.

This number is most impressive, but perhaps less impressive than it looks. Most of the plants are used in a minor, casual way. Almost any plant that seems astringent is used as a wash for skin conditions. Almost any small vine is used as a wash for *ojo* (on which see below). Almost any tannin-rich tree bark is made into a tea for diarrhea. Any wild tuber can be mashed and put on wounds and sores. In short, assigning medical value to a plant is often done on the basis of very general qualities, and the healer simply uses the first plant encountered that satisfies these criteria. As one healer put it: "Sometimes one plant is available, sometimes another. If you can't find the best one, you use whatever you find." Even so, simply knowing the astringency level or tuberous habit of 500 kinds of plants is already a rather major accomplishment.

Almost all plants used are flowering ones. Nonflowering plants are very poorly cognized by the Maya. Algae, fungi, and ferns are used very sparingly, and no specific names for them are used. The only native gymnosperm (a cycad, *Zamia lodigesii*) is used, but mainly for rat poison.

Among flowering plants, 89 families contribute medical herbs. The usual suspects—the mint family, bean family, and spurge family, among others—contribute a disproportionate number of drugs, and the grass family and orchid family contribute disproportionately few, as elsewhere in the world. Also fairly typical in worldwide perspective are the popularity of the Rubiaceae and Rutaceae. Moerman (1998) has shown that North American Native peoples draw disproportionately on certain families for drugs. The Maya are typical in their fondness for the mint and bean families. Perhaps less typical is the relatively high usage of certain minor families, such as the dogbane family (Apocynaceae). Especially useful is the verbena family. Almost every Verbenacea in the state seems to be a cure, and many are important, widely known ones. Composites (Asteraceae) are widely and heavily used, but not disproportionately so, given their abundance (both number of taxa and number of individuals being huge).

Robert Voeks (1997) has noted that most plants used medicinally in coastal Brazil are garden, weedy, and second-growth plants, not plants of the deep forest. This is true for the Maya as well, but not to the same extent. The Maya use almost everything, wherever it occurs, except for plants so rare and obscure that they are not worth seeking. Many of these are deep forest plants, which makes the forest relatively less a source than it might be. The common and obvious forest trees, however, are all well used, and several of the most important drugs come from the forest; some are not common and must be sought. The natural tendency is to grab the nearest plant, which will, of course, generally be a garden plant or town weed. Counterbalancing this, though, is the great popularity of such deep-forest remedies as 'elemuuy, *wako*, and the various aromatic plants called *taankas-che'*. These are sought out carefully. They are a part of the Maya strategy for using the forest; everything is seen as valuable and usable, unless it is truly rare and obscure. This provides an incentive to conserve and manage the forest, which the Maya of west-central Quintana Roo do very well.

To be exact, 125 of the plants I have recorded are forest plants (some are sometimes cultivated too, notably sapote and palmetto). Another 122 are weedy (including short-lived trees that do not persist as forest regrows, but not trees that invade abandoned milpas as part of forest regrowth). Yet a further 104 are cultivated. Three are water plants. There are surprisingly few plants that cannot be easily classified into one of these categories. The most important is *Dorstenia contrayerva*, a major cure-all; it invades milpas abundantly, but persists in mature forest. I have logged it as weedy. Cedro can be a cultivated tree, a forest tree, or a weed, but is almost always seen as a cultivated tree in my area, as is so logged here. (The list of "weeds" of Yucatán given in Villaseñor Rios and Espinosa Garcia 1998:308–13 includes 108 of my plants, plus 20 further plants that are members of genera in my list.)

By comparison, the total flora of the area (insofar as I have identified it) includes 208 forest plants known and named by the Maya, 205 weedy ones, 235 cultivated plants (not counting wild plants often cultivated), and 4 water plants.

Maya medicine is a developing, productive science. Several new weeds have invaded western Quintana Roo in recent years. Many of them now have medical uses—inferred or constructed by expert healers. The ones used are aromatic or astringent (e.g., *Sonchus oleraceus, Conyza canadensis*), as opposed to things like grasses that are regarded as mere pests.

Medical knowledge can be conveniently divided into three sorts: remedies that are almost universally known among traditional families; remedies that are fairly common knowledge, but known only to people seriously concerned about herbal lore, usually senior male or female heads of households; and remedies that are definitely expert knowledge, known to and used by professional or semiprofessional healers. Such healers fall into two categories: jmeen, who are male herbalists and ritual officiants, and midwives. The jmeen know hundreds of plant species. Midwives know many fewer species, but their specialized knowledge of birth includes some special plant lore (Anderson et al. 2004).

The three most famous and distinctive medicines of the region are 'ele-muuy, *wako* (*Aristolochia maxima*), and *kaambajauj* (*Dorstenia contrayerva*). 'Elemuuy, as mentioned above, is basically a diuretic, used for any and all kidney and urinary conditions (and occasionally for other purposes). Wako goes most often into stomach preparations. Kaambajauj is a cure-all, used internally or externally for almost anything, from fever to snakebite. 'Elemuuy and wako are both deep-forest plants, though very rarely they are grown in gardens. Kaambajauj is a weedy perennial, not deliberately planted, but encouraged when it invades fields—as it generally does without any need for help. Another cure-all, taankasche', is rare and found only in certain parts of the deepest forest.

'Elemuuy, wako, and kaambajauj are all commercialized to some extent. Attempts to farm them have been made, but they are common enough in the wild to need no special attention so far. They are even brought as far afield as Los Angeles by Maya traveling there. (A Yucatec Maya beachhead exists in Inglewood, a suburb of Los Angeles; Maya foods and remedies are available there if someone has recently been visiting home.)

Though these plants are seen as the high-line drugs, recourse is more often made to garden plants and weeds, since those are readily available. Virtually every family grows medicinal herbs in the garden. Most of these are Spanish introductions: mints (*Mentha* spp.), rue (*Ruta chalepensis*), aloe vera (*Aloe* cf. *barbadensis*), citrus fruit trees, etc. The uses came with the plants, including not only obvious matters like using mint for stomach upsets, but also the use of rue in magical practices—an old Spanish tradition widely established in Mexico.

Weeds, however, are a different story; they are almost all native plants used in traditional Maya ways. *K'anaan* (*Hamelia patens*) is one of the commonest, being not only a universal weed but also planted in many gardens for medicinal use. It is the most popular antiseptic wash, being used on all skin conditions. It appears to be effective. *Jawaij* (*Parthenium hysterophorus*, Asteraceae) is popular for a tea for respiratory and other ills. It grows on every roadside, so few bother to grow it deliberately. A range of small weedy fabaceous vines make washes for the effects of the evil eye (ojo).

As elsewhere in Latin America, contemporary Maya medicine depends heavily on maintaining a balance in the body between heating and cooling influences (Foster 1994; Lopez Austin 1988; Ortiz de Montellano 1990; for the Maya, McCullough 1973; Redfield and Redfield 1940; Redfield and Villa Rojas 1934:161–64). For the Maya, the greatest danger is encountering cold when overheated. Most diseases are caused by drinking a cold drink, or encountering a cold draft, when one is very hot from work or fever or the sun. Cures, therefore, often seek to cool the overheated (using cool baths for fever is especially favored) or to warm the chilled (honey and spices, taken internally, are most favored for this). However, more often, people use herbs that are known to be effective specifics for the symptoms at hand, whether or not they are "heating" or "cooling." Thus, for example, guava bark is used for diarrhea not because it is "heating" but because it works. (It is well known in international biomedicine, and used clinically in tropical countries all over the world; Barbara Anderson and E. N. Anderson, personal observations.) Mint tea, similarly, is not apt to be considered "heating" when it treats stomach conditions (almost invariably considered to result from cold drinks); it is correctly regarded as simply a stomach soother.

Herbal cures are the usual ones for all such conditions, as well as for bites, skin problems, and indeed almost all diseases. Many of the herbs are known to work (Ankli 2000). Nonindigenous observers may try to separate herbs that "work" from herbs that are "just magical." But the leading magical herbs seem to have empirical effects; *ruda* (rue, *Ruta chalepensis*), introduced by the Spanish, is an effective stomach medicine, and its native equivalent, siipche' (*Bunchosia glandulosa*, a small tree), has real antibiotic effect (Ankli 2000). Otherwise, herbs are used primarily for pragmatic, naturalistic medication. They are believed to work, simply because experience has shown that they do. Some are believed to work only because they resemble others that do, or because they look as if they should (bright yellow indicates a jaundice). Working elsewhere in the peninsula, Anita Ankli found good agreement among healers and found that they expect medicinally useful plants to have a strong smell or taste, especially pungent or bitter (Ankli 2000; Ankli et al. 1999a, 1999b). This is not a major concern in Quintana Roo, but similar beliefs exist. She found the theory of heating and cooling medicines to be more important in evaluating herbal cures than it is in Quintana Roo.

Antiseptic action has been demonstrated in laboratory studies of siipche', k'anaan, guava and nance bark, and many other plant remedies. Worldwide remedies like mint, rue, aloe vera, wormwood, and epazote are well known. Revealing further details on other known cures must, unfortunately, await a day when better protection is afforded for indigenous intellectual property rights. My experience suggests that most Maya internal cures, and many of the skin disease washes, work to some extent. Some work quite dramatically, better than drugstore remedies. Most work less dramatically, and sometimes barely perceptibly. Within recent memory, clinics were far away, and one was often isolated in the forest. At such times, anything was better than nothing. Even today, the clinic is often closed or distant, and so the old cures persist—even the barely functional ones. Finally, herbal baths for conditions like malaria and insanity clearly fall into the category of desperate hope-against-hope; few, if any, Maya are surprised when these remedies fail in a particular case.

Herbal teas, both drunk and used for washing and bathing, remain the commonest remedies for both natural and supernatural diseases. Supernatural conditions also require a brushdown with a bundle of ruda or siipche' twigs.

Magical and supernaturally caused illnesses require a ceremony—normally carried out by a jmeen—to cure. Probably the commonest is the evil eye. Pregnant women, drunks, overheated workingmen, and a few other types of people supposedly have the evil eye and thus can cause diarrhea in children by simply looking at them. Occasionally this is thought to be done deliberately, but the vast majority of cases are inadvertent. This condition is treated by herbal baths, whereas ordinary diarrhea is treated by herbal teas. Normally, evil eye is diagnosed by elimination; if the ordinary cures do not work, evil eye is suspected. Most of the ordinary herbal teas are known to be effective by international biomedical standards (Ankli 2000), so there is indeed something to explain when they fail. The baths are not so effective, but they reduce fever and discomfort.

Witchcraft and sorcery—*brujería*, deliberate working of malice—is much less common, but much more serious. Also, deities and spirits affect people. Some of these are malevolent; some are only doing their job, sending disease as punishment for sin. More common are supernatural winds or spirits that mean neither good nor evil, but are simply *k'aak'as 'iik'* ("bad winds"). Everything from lightning bolts to the demon opossum (*bok'ol ooch*) can affect people with mental illness, general wasting, or sheer bad luck—not intentionally, but simply by being what they are. Magical and supernatural affections are almost always treated by herbal baths, and, in serious cases, by ceremonies also.

Magical ailments include some types of nervous attacks. An occasional problem, which can be due to witchcraft or to psychological state, is dislocation of the *tipte'*, the lump under the navel; it is the center of the body, and must be kept in the right place to maintain health. (See also Hanks 1990. This belief is widespread in rural Mexico.)

Rituals for magical or spirit-sent illnesses range from simple herb teas through a *limpia* (brushdown with minor religious ritual) to a full-scale *kex*, a cleansing and healing ceremony that can be quite complex (cf. Faust 1998). A classic source on ancient Maya medicine is the *Rituals of the Bacabs* (Arzápalo Marin 1987; Roys 1965), a sixteenth-century book written in Maya using Spanish letters. It includes sixty chants for treating supernatural conditions, especially *tancaz* (*taankas*, now "paralysis" or "numbness," but then meaning a wide range of mental illnesses).

Considerable Spanish magic and spiritualistic lore has invaded Maya medicine. The degree to which it has penetrated the Maya world differs greatly with individuals. Don José has diligently sought it out, learning all he can about it—and also about modern medicine; his cures may involve

Maya herbs, Spanish magical practices, local patent medicines, and drug-store antibiotics, all at once! At the other extreme, Don Marcos's healing is purely from Maya tradition; he knows and uses only the most universally known of Spanish magical practices. One of these is the need to gather certain ritually used herbs on Tuesday and Friday, the bad days of the Spanish religious calendar; they were, in Roman paganism, the days of those troublemakers Mars and Venus, and as such were highly suspect to the early Catholic church.

This brings us to the question of the future of Maya medicine. The herbal lore of the Yucatec Maya has been documented now in many sources. Much remains to be done, but the general contours of the system are clear, and a huge range of drugs has been listed. Unlike many indigenous cultural traditions, Maya medicine shows no signs of dying out. However, the eso-teric knowledge of the jmeen is not always being passed on.

The following issues are being raised in Maya circles.

Commercialization. Farming the major medicinal plants has been tried. So far this has been unsuccessful, because there is no market, due to the ease with which plants can still be gathered from the wild. As population increases and forest shrinks, this will change, and farming will become profitable. It will probably start in door-yard gardens and spread from there. Limited commercialization of dooryard medicinals has already happened, and is increasing. The age-old tradition of swapping garden specialties between families (cf. Herrera Castro 1994) has provided a basis or skill development arena for local marketing.

Further research. So far, the work of Ankli and her collaborators is the only systematic research on the actual value of traditional Maya medicines, though many of the individual plants are widely known and have been tested and found valuable elsewhere (e.g., guava, mint).

Intellectual property rights. Insofar as valuable healing effects are found in Maya plants, how can we get these to the world without cheating the Maya? International drug companies are now wary of commercializing traditional remedies, because of fights over intellectual property rights. The Maya case is particularly vexed, because the Yucatec Maya are a cultural and linguistic unity but are not a political or social unity. Moreover, their medical lore is not the same everywhere. In fact, it varies enormously, according to local vegetation, ecology, and cultural tradition. There is thus no way of assigning a particular remedy to "the Yucatec" or to any one community. I have encountered remedies that work far better than any biomedical drug. These remedies are known to experts in several communities; they are not the property of one person or of one community or of the Yucatec as a whole. How could we possibly assign rights or fairly distribute benefits if we commercialized these remedies, which the world seriously needs? Any solution that occurs to me seems unfair. One could set up a Maya medical corporation that could distribute benefits widely or invest them for the benefit of all, but how would one decide who should belong? One could simply give the benefits to the Mexican state, but that polity has, alas, spent much of its history fighting the Maya in wars of oppression—a point that the Maya rarely fail to make when discussing issues like this one. One could attempt to identify all the experts who know the remedies, and compensate them, but the practical difficulties of doing this are insuperable. (People learn faster than bureaucrats can find them.)

The "etic" side of the medical situation may now be introduced. This is what the local doctors say when they use the discourse of modern international biomedicine.

Spraying, drainage, and treatment essentially eliminated malaria years ago. Migrants occasionally reintroduce it to Quintana Roo, and there is always the danger of new outbreaks, but rigorous steps keep at bay this deadliest of the old-time scourges. Immunization in the rural clinics has controlled other infectious diseases. Inoculations have not only immunized people; they have broken the cycles of transmission of the diseases in question, which have consequently died out.

Minor diseases are common, due to poor sanitation and unclean water. Minor respiratory ailments are by far the commonest medical problems (INEGI 2000c:79f). Ascariasis is the most commonly reported disease in the local government clinics. Amoebic dysentery, giardiasis, pinworms, and other parasites come close behind. Parasites are rarer than they used to be. Intestinal ailments remain common; *Giardia lamblia* is pandemic in the water supply, *Salmonella* spp. are common, and various minor intestinal complaints affect everyone sooner or later (data from interviews with local medical personnel). Malaria and yellow fever have been eliminated, but dengue, epidemic in Yucatán state in recent years, has entered the area; Chunhuhub recorded its first cases (in recent history, at least) in 2002.

Virtually everyone suffers from at least one bout of disease per year, and children especially tend to get sick. However, the town is generally healthy. (The only illness worthy of note that I got in the peninsula was a case of salmonellosis I picked up in a fancy restaurant in Mérida. A year of eating everything in Chunhuhub never even gave me indigestion. The local inhabitants, more seasoned and immunocompetent than I in dealing with the area's microbes, stay correspondingly healthy; cf. Peraza Lopez 1986; Crooks 1997).

Skin infections and fungus are also universal in this tropical rural world. They are now easily controlled with drugstore medicines. However, older people (myself included) remember days in tropical America when the drugstore had nothing effective for fungus, and nothing really effective and safe for skin infections. The Maya used—and still use—a vast range of plant medicines, many of which are highly effective.

Violence is rare. No murders have occurred in many years. There was one suicide while I was in Chunhuhub in 1996: a young man, alcoholic and mentally ill.

From 1970 to 1990, Quintana Roo grew from about 88,000 to 493,605 people (INEGI 1987, 1991). The birth rate is currently around the national average of 21 per thousand per year, while the death rate is only 5. Quintana Roo grew in population by an average of 5.1 percent a year through the late 1990s (INEGI 2001:6), by far the fastest of any Mexican state. Naturally, it also had the highest percentage of population from other states, 56.1 percent in 2000 (INEGI 2001:20; most of the migrants were from nearby Yucatán and Campeche).

Chunhuhub continues to grow, because of a spectacularly high birth rate. Settlers have usually been young families, and Maya families are famous for their large size. Old people live to a considerable age. The demographic pyramid resembles a melted soft ice cream serving: a tall but exceedingly thin point above a widely spreading cone. Conditions in Chunhuhub are typical of those in Felipe Carrillo Puerto municipio. The municipio birth rate is 50 per thousand; women in the forty-five to forty-nine years of age bracket have, on average, 7.5 living children. (After that it declines because of accidental deaths of the young. In my figures as in INEGI's, completed families typically run seven to eight.) The death rate is 2.7 per thousand. (See INEGI 1994:20–21.) Thus Chunhuhub has grown from nothing in 1940 to 6,000 today, and at current rates of growth (4 percent for the municipio; INEGI 1994:15) it will be a city of a million people well before 2200. This is not a long time for a town with a two-thousand-year history. At the state's growth rate of 8.3 percent, Quintana Roo will, by then, have about as many people as the entire world does today.

The infant mortality rate was an incredibly low 9.4 (per thousand) in 1990, and has dropped since. (For some reason infant mortality increased in the municipio in 1997, from 11.8 in 1996 to 20.5 [INEGI 2000c:11]. This has corrected itself, and levels are down again.) Even granted that more than a few neonatal deaths are not brought to the statisticians' attention, the low infant mortality rate of Quintana Roo represents one of the most creditable accomplishments in modern history. It is far lower than the infant mortality rate in American inner cities, or in United States rural areas that approach Chunhuhub in economic conditions.

The reasons for the low rate appear to be twofold. First, and by far the most important, is Mexico's excellent health-care system. Chunhuhub's clinic is well staffed and well run, however much its doctors may feel deprived of city amenities. (One young doctor wistfully asked me if I played chess. I don't. No one else in town did, either.) Chunhuhub also has two other doctors, as well as competent pharmacists. These are backed up by excellent health-care facilities in Felipe Carrillo Puerto and Chetumal. Certainly, a person of humble means seeking medical care in Chunhuhub is better served than she would be in most parts of the modern United States. Second is the strong family system and excellent folk health care of the Maya. It has to rank second, because when the Maya had nothing else, there was that five hundred per thousand infant mortality rate; but, with modern inoculations and antibiotics to back it up, the strong family support system and the herbal tradition still provide good care and keep people alive and well.

Parteras still deliver 22.7 percent of the babies in Quintana Roo (vs. 17.1 in Yucatán and 21.1 percent nationally; INEGI 2001:119). They receive some government instruction (there are three courses a year in Quintana Roo; INEGI 2000a:163) and clinic support. Parteras oversee all pregnancies and most births in the Chunhuhub area, sending women to the hospital in Felipe Carrillo Puerto if problems arise or are predicted (Anderson et al. 2004).

A diet of maize, with relatively slight amounts of beans and meat, remains standard. It is not always adequate in vitamin A and minor B vitamins, but it is if much use is made of the local fruits and vegetables. However, it does not usually provide enough iron for women, especially women who are pregnant or nursing virtually all their adult lives. Clinics routinely prescribe iron supplements, but low-level anemia is rampant, and polymineral deficiency weakens many a new mother. As recently as 1998, malnutrition killed a recorded thirty-five people in Quintana Roo (INEGI 2000a:31); probably most of these were young children in thorn-forest ejidos, an environment where I have seen hungry children and have seen some photographs (taken by a now deceased doctor of Chunhuhub) of literally starving ones.

Vitamins prescribed are often imbalanced. Injections of the minor B vitamins are popular with local health agencies; the area needs complete multivitamin supplements with vitamin A and iron, not just B vitamins, though folic acid supplementation is necessary in many cases and a good idea in many others (research observations by Barbara Anderson and myself). Vitamin A deficiency is reported as having occurred in the past, though it can never have been common in this world of yellow corn, sweet potatoes, chaya, chiles, mangoes, and papayas.

There is some downright quackery; various proprietary compounds of little or no medical value, as well as vitamins hyped with valueless herbal supplements, find their way into the nutritional prescriptions locally popular. Maya pay inflated prices that they can ill afford for these and other medicines; cheap drugstores are not found in the countryside.

People are aware, to varying degrees, of nutritional values of various foods. Maize is believed by traditional people to be the perfect food—the basis of a healthy life. Honey is regarded as especially healthful. Fat is desirable. Meat and beans are known to give stamina and working ability. Conversely, long and bitter experience has taught all older people that one does not maintain weight or strength on famine foods (wild greens, root crops, and the like). They are low in energy and hard to digest. Thanks to the school system and the local doctors, concepts of vitamins are now well established among those with some education (including all the younger people), and many are aware that foods like chaya, mangoes, and papayas are high in them. Most are also aware of the dangers of untreated water and unclean food.

However, by far the biggest medical problem on the near horizon is the enormous increase in junk food consumption. It is becoming recognized, but little is done. Children very frequently have a sweet or salty snack in hand or mouth. Sodas are universal—the general sign of hospitality as well as the regular drink. Fresh fruit juice is beginning to catch on; the breakfast stalls in the market now routinely supply fresh-squeezed orange juice, a very new touch, and the *loncherias* are supplying a wider variety.

The results of the junk food epidemic are serious. The Maya seem genetically prone to obesity and diabetes. Obesity is now claimed to affect 45 percent of the men and 75 percent of the women in Yucatán (*Diario de Yucatán*, September 24, 2001: "Que la obesidad ya es una epidemia en Yucatán"). I believe these figures are exaggerated, but the reality is certainly uncheering, and Quintana Roo is not much less overweight than Yucatán.

Diabetes is now the third most important cause of death in Mexico, with 36,027 deaths in 1997—a long way behind cancer with 51,254, but growing fast; diabetes killed only 27,139 in 1991 (INEGI 2000d:15). Today, more than 5 five million Mexicans suffer from it (*Por Esto!* September 21, 2001: "Cinco Millones con Diabetes"). Diabetes killed 158 in Quintana Roo in 1998 (INEGI 2000a:31). The indigenous people of Mexico, including the Maya, have extremely high levels of the genes that predispose individuals to adult-onset diabetes. They may also be genetically programmed to gain weight easily (the so-called thrifty genotype). Sugary and floury snacks contribute to the high level of obesity in the population. Diabetes levels are skyrocketing among the Maya, including those of Chunhuhub. The problem is now well recognized, and advice is beginning to turn against junk food.

Medicine and medical care, for the contemporary Maya of the Chun-huhub area, form a complex and closely woven tapestry. People maintain health, and treat the endless minor problems and occasional major ones that compromise personal and family viability, in a world where several medical alternatives vie for their attention. In particular, they are aware of a profound divide between the "modern" medicine provided by the local clinic and trained doctors, and the vast and diffuse web of traditional home remedies, religious healing, and specialized herbal and curing knowledge of indigenous curers. They do not, however, separate these in practice. Pre-eminently pragmatic in this as in all things, they combine all the above approaches, according to what works (or seems to work, or is said to be capable of possibly working) in a given situation.

CHAPTER EIGHT

A Few Conclusions

ON THE SMALLEST and most simple and direct level, the world can learn a great deal from such experiments as the Sosa farm, or the milpas being developed in Sahcaba and elsewhere using velvetbean as green manure (Mizrahi et al. 1996), or the integrated development modeled by Tres Garantias and Nohbec, or the connection from conservation forestry with skilled guitar makers (Canopy 1998), or the honey projects scattered throughout the peninsula (de Jong 1999). The Sosa family's intensive vegetable and fruit garden, the ecotourism initiatives in Tres Garantias and Sian Ka'an, the archaeological tourism of Coba and Dzibanche, the peccary farm at Ake, the production of embroidered clothing and table-cloths in Xpichil, and dozens of other modest but successful projects also feed into this future. Farther afield, peccaries, deer, agoutis, pacas, and crocodiles are all being successfully farmed in Mexico or other parts of Central America. Small-scale yet intensive fruit production—not just of citrus but of much more valuable tropical fruits such as cherimoya and mamey—flourishes in many parts of the Caribbean basin.

There are dozens of such small and hopeful projects all over southeast Mexico. All of them should be receiving far more attention. The world spends trillions on armaments, typically used to seize and waste other people's resources; hundreds of billions trying to alleviate the poverty and disease caused by environmental abuse; and perhaps a few million trying to alleviate environmental abuse itself.

By adding together such small and unobtrusive initiatives, rural Quintana Roo could prosper. But there are more exciting possibilities within reach. Chunhuhub has carved out a leadership position in the world of computer training in Quintana Roo. At present, graduates of CEBETA must go to the cities to find work. In the near future, however, that will not be necessary. With the coming of reliable electric and telephone service, Chunhuhub could become a leading center of all types of computer-based activities—an "electronic suburb," a Durango or Aspen of south Mexico. The young people of Chunhuhub (like the youth of suburban California) show a remarkable affinity for these machines. There is every reason to keep them at home, where they can continue to be a part of the family enterprise. Nothing is more traditional and long-established among the Maya than multiple tasks in one compound. There is every reason to expect that a family could combine (for example) software development, milpa agriculture, specialized fruit production, technical services, and beekeeping. Combinations equally "improbable" are already common in Chunhuhub. (Recall the family compound that unites intensive vegetable culture, hair-cutting, newspaper delivery, schoolteaching, religious ministry, and other activities.)

Some such development is necessary to hold the "best and brightest" in the town. Otherwise, Chunhuhub may suffer the fate of so many farming communities: decline into stagnation, or even social pathology, because its most dynamic and intelligent people are gone. In a community as "public" and as network-organized as Chunhuhub, with few firm institutions but a strong and vibrant grassroots tradition, this is a matter of particularly serious moment.

The Maya have a tremendous wealth of knowledge about their biotic environment—the fields, forests, orchards, and waters from which they make their living. Some of their beliefs (such as the efficacy of witchcraft) do not stand up under independent empirical testing, but most of their knowledge is grounded in—and stays close to—hard-headed, verifiable, empirical observation. It is thus reasonably described as "science" (cf. Anderson 2000; Gonzalez 2001).

A system of thought should be understood in its own terms, as a united whole. This does not mean that it is a *homogeneous* whole. All systems have their ambiguities and debated points. The referential system that Maya use when working with and talking about plants and animals is a genuinely shared system; people understand each other when they talk about planting, harvesting, hunting, and house building in terms of the Mayan cosmovision. But this is not a written system with textbooks and drills. It exists only in Maya practice (Hanks 1990). The Maya (perhaps more than some other people) are individuals, even individualists; they love nothing better than a good rousing debate over the fine points of hunting deer, or over the proper name of a tiny insignificant bird, or over the proper way to cook chaya. The point is not that they agree on everything; *the point is that they agree on the terms of the debate.* They can debate because they share a cosmovision or frame of reference, based on practice and continually defined and redefined in terms of that practice (Hanks 1990; William Hanks's discussion of this matter in his analysis of Mayan referential linguistics is so detailed, accurate, and superb that I can only refer readers to his book; it is long and difficult, but few if any works are more rewarding not only to Mayanists but also to the anthropologist, the linguist, and the philosopher of science).

Maya traditional biology has a great deal to offer. Maya agronomy includes a great deal of knowledge that is valuable on a world scale. One need think only of the plants that seem likely to be Maya cultigens: chaya, mamey, and *k'aniste'* fruit, and others. These could revolutionize farming in some parts of the world. Maya folk medicine is also an almost untapped resource. Testing of its value has only begun. Adequate study will surely be repaid with new and improved remedies for many conditions. Maya forestry embodies knowledge that is necessary for optimal management of Quintana Roo's endangered timber resource base.

Far more important is the ability of the Maya to manage resources, collectively, for the long term. This contrasts with the failure of so many other groups to do so in the tropics (Atran et al. 1999; cf. Terborgh 1999) and elsewhere. It provides a thought-provoking case with implications for the trend, now widespread, to bring the successes of traditional resource management to the world's current conservation problems (Laird 2002; Stevens 1997).

Ecological anthropologists are aware that one cannot simply pluck "useful facts" from a traditional system of knowledge, insert them in the pages of a learned journal, and be done. The facts exist in a context of traditional beliefs. They cannot be fully understood apart from this context. Pulling out of context a "useful fact," the agronomist or doctor may lose

most of the relevant information—not just about method of use or necessary steps in the procedure, but about the motivation to use the bit of
knowledge in question, the proper situation in which to use it, the economic and psychological context, the emotions and rituals that keep the
"fact" salient and available.

On the other hand, plants and medicines, methods and techniques,
have to be taken from their full indigenous context if they are to be of
any use to people around the world. I have no apologies about the chaya
and *mak'olan* plants growing in my backyard, or, for that matter, about the
(equally Mexican) squash and sunflowers there. Chaya, to the Yucatec, is
not just a food plant. It is associated with the *Xtabay* (the demon woman)
and other strange beings; it has cultural and personal associations bound
up in its role as a famine food, an animal fodder, a marker of indigenous-
ness. All these matters must be understood by an anthropologist studying
the place of chaya in Yucatec culture. However, chaya is, preeminently, a
food. In fact, it is a food plant virtually unique in the world in the degree
to which it combines tolerance for harsh conditions with high nutritional
value. Such a plant is desperately needed by literally billions of people,
on every continent outside of Antarctica. Propagating it worldwide as a
nutritional godsend should be done immediately. It need not, and should
not, be mixed up with the Xtabay and other matters.

More important, and far less amenable to uprooting from its home culture, is the Maya attitude toward the biotic world. This point has recently
been made by many thoughtful Maya, and by many thoughtful Mayanists
(e.g., de Jong 1999; Faust 1998; Lenkersdorf 1996). On the whole, traditional Maya view plants and animals with respect, concern, care, and
even love. They enjoy watching birds and insects. They love being in the
forest. They care for and manage biotic resources with tenderness and
regard. They are not naive "greens" or "animal rights" activists; they have
a living to make, and they have to use plants and animals to survive. They
know they have to kill some animals, and they know that some animals
(jaguars, fer-de-lances) have no compunctions about returning the favor.
Their attitude of caring and concern is that of a participant, a manager, a
member of the web—not that of a park keeper. It is not a preservationist
ethic, and therefore perhaps not ideal for managing the core areas of the
new reserves that have been established to preserve indigenous fauna and

flora. On the other hand, one remembers that these areas have all been subjected, over the millennia, of Maya slash-and-burn agriculture. They are, to a great extent, the products of Maya management. Thus, even in strictly preservationist activities, we ignore Maya tradition at our peril. This is perhaps especially true in rural Mexico, an area in which the nonindigenous population has until recently had a highly negative attitude toward the forest and the wild—seeing naturaleza as something to be destroyed to make way for civilización.

On the other hand, we cannot say that the Maya cosmovision is the be-all and end-all for the whole world. It is surely not transferable in its entirety; Asian and African cultivators are not going to worship the Yuntsiloob. Moreover, it has its limits. Traditional Maya controls on hunting (even where they endure) have fallen hopelessly behind the realities of population pressure and widespread gun ownership.

Use of indigenous knowledge raises serious questions of intellectual property rights. Who should be compensated if the world adopts Yucatec Maya management strategies? The Yucatec are not a corporate group; one cannot tell who is and who is not Maya. When I discuss such issues with my Maya friends, they invariably say that their knowledge should be out there for all humanity, without charge—except that they expect full protection for, and security of, their land rights! This is a noble attitude, but it will not satisfy the lawyers, and the issue remains daunting (Anderson 2003; Brown 2003).

The question of saving and building on indigenous rights raises an even more ferociously difficult moral issue (Brown 1998, 2003). There are questions such as, who are indigenous people? and, how do you determine what is indigenous? Of more general interest is the age-old question of individual versus collective rights. How much do we treat indigenous people as individuals, equal to other members of the nation-states in which they find themselves? By contrast, how much are they to be treated as collectives—"the Maya" or "the Zapotec"? Are they nations with rights, or groups of people with nothing more than particular—and often recently constructed—identities?

One would think that the old-fashioned structural-functionalist view would go with a view of indigenous people as true collectives, with collective rights and privileges, including the right of collective self-determination about land use. Conversely, those who see ecosystems and societies as mere illusions, and see the world solely in terms of individuals maximizing their short-term advantage, should be expected to see indigenous peoples as mere collections of individuals, with no special rights. On the whole, this is true, but there are exceptions.

The hardest problem comes when people invoke a communitarian ethic and do indeed see indigenous peoples as collectives with particular rights and privileges. For, when this is done, the nations in which these people find themselves almost always invoke the same communitarian ethic to excuse forced conformity, and even forced acculturation ("culturocide"). If it is desirable for the X to be seen as a unit, with their own sacred traditions and ways and their own privileged access to land and resources, how can the nation of whom the X are citizens (and only a small percentage of the citizens) be denied the same privilege? If it is good to maintain cultural conformity, how can we intervene to stop nations from forcing conformity on unwilling minorities? Indeed, how can we protest apartheid or outright genocide? The extreme communitarianism of certain philosophers, such as Alasdair MacIntyre (1988), is—by any fair reading—downright supportive of religious persecution and easily bent to support genocide; yet it is the same philosophy that is invoked by anthropologists in at least some of their defenses of indigenous rights. Conversely, the extreme individualism of some libertarians and old-fashioned liberals would return us to a state of Hobbesian "warre," not least in the scramble for the environment that it would release. As with social and ecological theory, a reasonable middle ground is fortunately becoming prevalent, but it does not resolve key questions (see Brown 1998, with comments; Brown 2003).

Questions that emerge from this problematic matrix include the future of the ejido: is it too communal, a drag on individuals? And what of economic justice—a sore point with many Maya? Does the moral value of redistributing wealth make up for the economic and personal harm that seems sadly predictable in redistribution schemes?

All this provides contemporary developers and change agents with a really serious moral problem—that is, one that involves tradeoffs between several unquestioned moral goals. We have to preserve biodiversity. We have to respect and defend individual rights, including the rights of people who may be repressed and exploited by traditional systems. Recall Lenkersdorf's observations on the dark side of communitarian values. We have to recognize the existence of communities, cultures, and societies. We are at least sometimes under moral and legal obligations (e.g., as entailed by treaties) to respect the rights of indigenous minorities and to treat them as collectives with particular entitlements. Granted that—as usual—extreme positions are indefensible, we are left with a literally infinite number of possible accommodations of these views. Any one of these infinitely numerous accommodations will inevitably look good to someone and thoroughly bad to someone else. One suspects that the philosophers will be debating these issues until the human race is extinct.

The scholar—Maya or non-Maya—will wish to record Maya cosmovision in as complete and accurate a form as possible, for many reasons—above all, to preserve it for future Maya. The scholar, ideally, will also want to bring chaya and mameys to the world's attention as sources of nutrition. The scholar, and every other moral human, will most certainly wish to preserve Maya rights of self-determination and rights to land. But the scholar may also wish to go beyond all these matters, and to see how the Maya way of living can be integrated into a world future that will be increasingly crowded, resource poor, and stressful. It is all too easy to imagine the future if we do not respond to its challenges. There is a very real chance that humans will not only exterminate themselves but will destroy all higher life forms in the process. At the very least, we must make tremendous efforts if we are to feed and clothe the world and to reverse the ominous trend toward more and more frequent mass murder, genocide, and civil warfare. We must, instead, work for a future in which the sum total of human wisdom and understanding is brought together to meet those challenges.

Chunhuhub is typical of millions of small towns the world over. It is strikingly reminiscent of the small Indiana and Texas towns where my wife and I have our roots. We can recognize the United States of fifty years ago in the forest ejidos of today: the pigs in the streets, the cold-water bucket baths, the single telephone, the cheerful—or at least brave-fronted—endurance of people who know no other life than one of incredible hardship and poverty. Chunhuhub is not "underdeveloped," let alone "backward"; it has computers, satellite dishes, and new cars. Indeed, it is more fortunate than most Third World rural communities. It is a sober, thrifty, successful town. However, one need only drift off the main road, to Presidente Juarez or Gavilanes or Santa María, to see genuine rural poverty.

Why does poverty persist in a rich world? Several canonical answers exist. They are all simple, plausible, and unconvincing (to paraphrase H. L. Mencken). If they were right, simple, and straightforward as they are, the world would have been fixed long ago.

First is the myth of the undeserving poor, which, interestingly, is probably the commonest theory in Chunhuhub (and in Indiana and Texas too): poverty persists because people are lazy, drunk, or tradition bound. There are those in the Zona Maya who drink themselves into poverty, but they are a small minority. Lazy people are also few; anyone really lazy would have starved to death long ago. So the theory holds little water. Many of its exponents in Chunhuhub are themselves desperately poor by the standards of the rich world. One can fully credit the unbelievable effort and sacrifice it took to get a few steps ahead, and duly respect those who did, even when they are snubbing their less thrifty neighbors. But one must also see that working like a slave for fifty years, and having nothing to show for it but regular meals instead of irregular ones, does not constitute an escape from poverty.

There are also countless families like the Dzibs of Presidente Juarez. They are hard working, thrifty, and sober. Their problem is that they have poor land in a poor community. Year after year they fail to raise even a decent crop. Without capital, and without opportunities, they are trapped. They cannot raise enough capital to get an education or start a business. They cannot even stay healthy; irregular eating of a poor diet, combined with lack of money for medicines and trips to the doctor, leads to frequent medical crises, many of which are extremely serious—and preventable for a few dollars. Sickness keeps them from working as much as they otherwise might, and so the cycle gets more vicious.

Second is the myth of the evil Wall Street devils: the United States, or some shadowy world conspiracy, has stacked the economic cards to appropriate Mexico's wealth on concessional terms. There is, of course, much truth in this. Blatantly unequal terms of trade have theoretically been replaced by "freer" trade, but the "free" trade in question is often a cruel joke. Typically, it pits multinational firms and industries that are actually heavily subsidized by the United States government against Mexican firms handicapped by their government's *burrocratismo* and *tortuguismo*. Worse: both United States and Mexican policies pit big firms, both alien and Mexican, against small firms and individuals. The big firms get full infrastructure provided to them; they are exempted (legally or not) from labor and environmental protection rules; they are often subsidized outright (Rosenberg 2003; Stiglitz 2003). They are given, in short, red-carpet treatment. The small operators have no such advantages; in fact, they are subject to discriminatory policies. This is exactly the system that Adam Smith was fighting *against*, back in the late eighteenth century (Smith 1910 [1776]). His name is invoked in defense of the very system he attacked. Free trade, for him, meant a level playing field, with both big and small firms free to find their own way—neither subsidized nor handicapped. This policy might well work. It is certainly not being tried.

Instead, Mexican small operators enjoy "free" competition that is as fair as a boxing match between Mike Tyson and a ninety-pound weakling with his hands tied.

For a notable and relevant example: United States maize and sugar, production of which is subsidized by the federal government by tens of millions of dollars of "welfare for the rich," compete directly with Mexican subsistence-level farmers whose production has been savaged by ill-advised policies. Large-scale agribusiness-type producers in the United States reap vast subsidies (Bovard 1991), while Mexico's small-scale producers, such as the Maya, labor under every imaginable handicap up to and including army invasion and looting of their lands (as in Chiapas).

Yet, even the unfair competition with the United States is clearly not the whole cause of rural poverty. Europe and East Asia developed and grew rich in spite of Wall Street. In fact, Mexico's crude economic growth rate was as high as Taiwan's during most of the late twentieth century. Why did Mexico stay poor?

Population growth is another partial explanation. The enormously high birth rate of the mid-twentieth century simply ate up the gains. This is a simple point, but a true point, and an extremely important one.

Corruption—especially the export of capital to numbered Swiss bank accounts—explains much of the rest. Mexicans sometimes exaggerate the extent of corruption in their country, but no one can deny its existence.

Hostetler even argues that when the Maya take money earmarked for a development scheme and spend it on beer instead (cf. Hostettler 1992, 1993, 1996), they may be doing the best thing. Following Mandeville's cynical argument in favor of luxury in *The Fable of the Bees* (1924 [1732]), one can say that government money spent on beer is at least spent in Mexico on Mexican products. In fact, given the level of planning of most Mexican government programs, diverting the money from a government agricultural plan to good honest beer is a clear improvement in many cases. The people know this perfectly well, which is why they do it. Intelligent, well-administered plans rarely wind up with their funds diverted to the bar. Certainly, money earmarked for things like clearing rain forest for cattle, or providing pesticides and herbicides to farmers, is infinitely better spent on beer than on its intended goal. The major problem comes when the money leaves the Mexican economy entirely—usually for Swiss bank accounts or foreign investments.

A secondary problem occurs when a really good program is looted to death. As Hostetler noted, programs and funds that are looted are usually those that deserve it, but some good programs—notably including ecological enforcement, as well as infrastructure development—are often subverted, to the enormous cost of the economy. Similarly, when a really necessary or desirable law is subverted, the results cost the economy a great deal. Game law enforcement is the most obvious case in Quintana Roo.

Perhaps the biggest cost of corruption, however, is its effect on the culture of doing business. In Mexico, as elsewhere, whom you know is often more important than what you know (or do). Networking, contacts, connections, and schmoozing are all-important. Quality of work, especially of planning, is neglected; why develop good plans when the reality is strictly a matter of personal connections? The attitudes and practices in question spill over into honest business and professional relationships as well, debasing otherwise honorable activities. Even legitimate business becomes a matter of networking rather than quality production. One can trace the history of this back to medieval Spain (and feudal Europe in general), with its estates and feudal relations.

Corruption is the tip of the iceberg. The rest of the iceberg is partially visible now that Mexican politics has become more transparent, thanks to the coming of a real two-party system. The underwater ice includes those unfair global terms of trade. It also includes, in its frozen depths, a lack of knowledge of marketing, and thus a failure to take advantage of the trade system.

Above all else, it includes *lack of accountability*. Mexico's and the world's elites are not just corrupt—they are outside the law. Recent revelations about CEOs' practices in the United States have made this clear. No one need fear that the CEOs in question will receive much punishment for their crimes. The high officials of the World Bank and International Monetary Fund are even less likely to suffer for their failures (Stiglitz 2003; admittedly a book that is biased against those agencies, but accurate on this point). Even development workers on the ground are not often accountable for their failures (Dichter 2003).

From the international agencies down to the municipio agricultural office, those who design bad plans do not bear the consequences. Misplanning is often deliberately corrupt, but it is also often done by the most impeccably honest, responsible, and diligent bureaucrats and developers. Mexican officials are a more conscientious and honorable lot than they are sometimes said to be. The problem is that they can, and do, make plans for—say—a year-long project to teach handicrafts, or develop irrigation, without having to prove to anyone that the plans are practical and will deliver anything. In the nature of things, all plans tend to cost more and have more bugs than anyone initially expected. They also take longer to bear fruit. Only accountability can bring planners up against the hard reality of cost overruns and extra years needed (see excellent and detailed discussion of these issues in Dichter 2003; see also Ascher 1999).

Again from my experience, the vast majority of plans for development in rural Mexico never had a chance. Even when very well intentioned and well thought out, they were based on inflated expectations. Local people develop cynicism and plan-fatigue, as well reported by Ueli Hostettler (1993). They take the money and, at best, use it while it lasts, but do not follow through by taking the initiative to continue the plan. Then the planners blame the locals for the failure of the plan.

The fault is not with the locals. They know that many a plan is merely an election-year device, and many another is well intentioned but quite irrelevant to reality. Thus they treat the plans as milpas: things to be cropped for a year or two and then abandoned. Being no fools, they can tell such "milpa plans" from the successful plans that actually continue year after year. These latter they treat as orchards: something to be cultivated and tended for the long run. Orchards are, in fact, the prime example of such plans; local orchard aid and credit schemes have proved lasting and successful. Social security, local education, and clinics have similarly proved themselves. Reforestation is beginning to do so. None of these suffer much from cynical diversion of money to the bar.

The fault for bad planning is not usually with the individual planners. It lies with a system that does not have a working concept of accountability or cost-benefit calculation. This is a problem not limited to Mexico; it is a worldwide problem that, in the opinion of veteran development official Thomas Dichter, has made a failure of the whole development enterprise (Dichter 2003).

Most extreme are the genuine tragedies of big dams and cattle deserts—the results of plans so insane that they have ruined half of rural tropical Mexico.

I find a small and tender example more revealing: *cunicultura*. This delightful word means "rabbit raising." For some reason, Mexican bureaucrats love to promote small-scale production of rabbits, to provide meat and income for the small farmer. (And not only Mexicans; I have already referred to Dichter's story of his own experiences with it in Africa.) The fact that Mexican farmsteads abound in small livestock (chickens, turkeys, and pigs) is not considered. Neither is the fact that chickens and turkeys give eggs as well as meat, whereas rabbits do not. Neither is the fact that rabbits in the tropics succumb to countless diseases. There is also the point that rabbits are not popular meat animals in Mexico.

In fact, the slightest consideration would make it clear that there is no way for cunicultura to succeed. Yet, millions of dollars have been wasted on this idea, not just in Quintana Roo but all over Mexico. I do not believe any corruption was involved. Surely, the rabbit industry did not pay off the developers. It was just foolish planning that was never subjected to any follow-up analysis.

A final irony is that many Maya have taken enthusiastically to rabbit raising—but for pets, for adored playthings of children. Few of the Maya, gentle people that they are, would dream of eating anything so cute.

Multiply this by many other examples, at all levels, and one realizes that bad planning, bad execution, bad English teaching, bad marketing, and many other small-scale local bads are the ways that the systemic flaws in the world economy translate to the local level.

These could be fixed by accountability, but accountability depends on people having the power to force accountability on the bureaucrats, and on the powerful in general. Such power has, until recently, been notably lacking in rural Mexico—except in cases where the people could take power into their own hands through violent means, a strategy that has had its own costs. Quintana Roo was free for a century, thanks to ferocious Maya defense, but the price included endemic malaria, lack of education, desperate poverty and malnutrition, and other miseries. The Maya felt it was worth it, and who can doubt them? But they no longer feel that way. They now have enough real power to enjoy modern comforts *and* keep their land. In Chiapas and in Guatemala, all too close, this is not the case, and fighting continues between Mayans and governments.

Will the two-party system, and wider participation in the modern economy, allow Mexicans to force accountability on local businesses, on the government, and on the unfair and unfree global system that has been dishonestly sold to them as "free trade"? Only time will tell. One need, at present, is for wider imagination and less aguantando. People have to believe a better world is possible before they can act to bring it about. They have to know what is available, and they have to know they could have it with a little effort. Programs that work build such hope and faith; programs that fail breed disillusionment, alcoholism, and hopelessness.

From the point of view of its probable effect on the future of humanity, the worldwide rural environmental crisis appears to be by far the most serious and immediate threat facing our species. Yet this seems to be a crisis not even recognized. If it is not resolved, the world food situation, already precarious (Brown 1995), will become rapidly worse. Already, about one-fifth of humanity is malnourished, and perhaps half are at risk of serious hunger. The Maya are poor enough, and many of them are getting poorer; but the Maya of the Mexican Yucatán are among the more fortunate of Third World countryfolk. I have seen and studied far worse situations in Cambodia, Bangladesh, and China, as well as in Oaxaca and some other parts of Mexico itself. Worldwide, the plight of the rural poor—still close to a majority of the human race—is extremely unfortunate. It is getting rapidly worse for many. In the future, when land, cheap fuel, forests, and fresh water are scarcer, the plight of the countryfolk will be far worse—unless action is taken now.

By contrast, other, more publicized crises seem less immediate. All-out nuclear world war would probably (not certainly) be more devastating, but the terror of "mutually assured destruction" is currently acting as an effective deterrent. Urban pollution is serious, and might in future endanger the human species, but so far it is less damaging than the multiple impacts that are slowly destroying the world's rural ecosystems.

Saddest of all, no one has really benefited from most of this waste of Third World resources. A forest area cleared, burned, and abandoned does not even enrich a gouging timber lord. It merely destroys trees. The animal life has been wiped out by casual shooting and pet keeping; no one got rich from that. A few giant agricultural firms have made money from overexploiting Mexican topsoil, but only in the short run; even the big firms suffer—and the small farmers and landless workers most certainly suffer—in the long run. One is constantly reminded of Clifford Geertz's bitterly accurate assessment of the history of Indonesia: "The real tragedy . . . is not that the peasantry suffered. It suffered much worse elsewhere, and, if one surveys the miseries of the submerged classes . . . generally, it may even seem to have gotten off relatively lightly. The tragedy is that it suffered for nothing" (Geertz 1963:143).

At least Yucatán, being flat, is spared from one of the horrors of rural overcrowding; in highland Mexico, people have been pushed onto steeper and steeper lands, and the consequent flooding, gullying, and general soil erosion devastates not only the steep fields but also all fields downstream. My ancestors in the Appalachians had a joke about having to hold onto a tree with one hand while cultivating with the other. My colleague Michael Kearney has photographed a man doing literally that in the mountains of Oaxaca. The poor man, a Mixtec farmer, had been pushed by land shortage to cultivate a piece of land as steep as a pitched roof.

However, the worst damage to planetary food potential has been caused by modern practices that lower overall productivity and are simply not sustainable even in the short run and even at constant populations.

Robert Redfield has often been accused, and not without justice, of contributing to the problem by his naive and simplistic model of modernization. He made traditional Maya ways seem all bad, modern urban ones seem all good. One may doubt if he really meant that, but he was certainly interpreted by many planners—in Mexico and elsewhere—as supporting such a position. Yet modern practices have failed over millions of acres in south Mexico, leaving wastelands. A clever but evil person is canny enough to avoid killing geese that lay golden eggs. Well-meaning innocents are not so clever. They not only fail to notice the golden eggs; they often do not even eat the meat when they kill the goose.

The first and last message of every "critique of development" is usually the same: there is a lack of fruitful communication between planners and the people on the ground—an effect of the lack of accountability and recourse. A range of opinion converges on this conclusion, from Arturo Escobar's "critique of development" grounded in political economy and political ecology (Escobar 1995, 1999) and Dichter's insider assessment (2003) to Graham Hancock's blistering independent-conservative critique (1991) and James Scott's broadside attack on the ideology and aesthetics of top-down, center-outward modernization (1998). Planners make decisions that are simple to make at a high and abstract level. People on the ground have no opportunity to comment and no recourse. In such cases, the planner does not look to see what the people on the ground might need, and the people on the ground have no way of telling the planner what they want. Some sort of communication between the two is obviously to be desired. This is what was different and special about the cases of Naranjal Poniente and Tres Garantias: planners and villagers actually listened to each other.

Inevitably, what they produced did not look quite like what either side had originally thought they wanted. The best thing about dialogue is that it can, sometimes, miraculously produce a better option than either side alone could have envisioned even in its best dreams.

Many anthropologists hold to a nostalgic dream of saving indigenous ways of life. I too hope for maximal preservation and minimal disruption, and so do many Maya—including the vast majority (I believe—from all I know) of those my age and older. But, realistically, the Maya are changing. They have been changing for thousands of years, and they do not want to stop now. They are acutely aware of the importance of computers, modern medicine, telephones, plumbing, and electric lights. Moreover, with higher population density, the traditional system of agriculture is falling behind. Overdriving it in a desperate attempt to feed the growing population merely leads to system failure and crash.

Only when both local communities and wider government agencies reinforce each other, and keep each other reasonably honest and accountable, can optimal solutions can be found. The state's authority cannot be maintained if everyone ignores it, or if its own enforcers believe so little in their work that they can be easily corrupted—or swayed by pleas of subsistence. The general public—from which, necessarily, the enforcement personnel comes—must believe that the goal is both worthy and possible. This, at present, requires that the general public have something like a traditional Maya attitude toward the natural world.

In short, in the twenty-first century, conservation can be effected only if five conditions are met:

First, the public must be concerned. They must care about the resources and want to manage them.

Second, they must have some way of getting together and maintaining social institutions that provide motivation. Experience teaches that economic concerns are not enough; there must be emotional, religious, or ideological involvement (Anderson 1996a).

Third, the government must be concerned enough to take recourse, accountability, and enforcement seriously. Latin America's long history of unenforced laws and "paper parks" must come to and end (Alvarez del Toro 1985). The public must see that the government will actually back up popular desires for sustainable management.

Fourth, there must be trained biologists and educators, willing to be planners, and, when necessary, whistle-blowers.

Fifth, these trained personnel must work with local communities and the general public. Biologists who talk only to other biologists, or even to other educated urban people, have some impact on general levels of awareness, but they accomplish nothing on the ground. It is all too easy to examine every side of a problem indefinitely, never coming to a conclusion, let alone acting on one.

Put in summary form: people must care about the common good and about the environment, and be responsible for those. Government must plan and provide accountability and recourse.

Lack of these conditions explains the failures of conservation in the tropics. Books, television programs, and political demonstrations have a major effect, largely on the first of these points. But, in the end, they accomplish nothing if they do not deal directly with the situation on the ground.

From the wider viewpoint of world resource management, it seems probable that the most directly and immediately effective way to save the world's natural resources is to *train biologists, educators, and other professionals, and include in their training the knowledge of how to work with local communities.* The latter sort of training is routine for development workers, such as agricultural extension agents, as well as public health workers. It should be mandatory for biologists, at least for those working in field biology, ecology, wildlife and forest management, and related fields.

The old gods may still be out there, somewhere; the properly attuned ear can hear whistling in the deep forest. The future, however, belongs to a new generation. The Lords of the Forest, in very human form, face the hardest challenge they—we—have ever met.

Afterword

ON JULY 18, 2004, Chunhuhub voted to privatize—to divide its land into individual parcels that can be sold. The ejido will survive, but its only reason for existence—land use and management planning—will be gone.

The vote took place in a fateful asamblea that lasted five full hours. A pleasant young lady from Mexico's agrarian bureau read the eighty-five articles of the code for privatized ejidos. Council president Alfredo Mo Pat and another council member alternated the reading, and translated as needed into Maya. Most of the articles are pure form and called forth no debate, but many required fine-tuning for Chunhuhub. These inspired a great deal of *wolyech'*—everyone talking loudly at once, everyone pressing on everyone else. The Maya normally share the quiet reserve and low-key but serious talking style for which Native Americans are stereotypically famous, but asambleas are a different matter, and one is not surprised to find that the Maya have a word for this concept.

The final vote, however, was no unequivocal victory for the forces of private property. The vote was around forty-two to one; more people had their hands up at the start, but several took them down again, leaving forty-two at the end. The lone stalwart opponent argued that the whole thing was badly planned and that everyone needed more time to deal with questions of fairness, including ways of dealing with nonfarming ejido members. Around 130 ejidatarios were at the asamblea, and many more had wandered in without staying for the vote; the total ejido membership is 330. So the vast majority were neutral. Debate showed they wanted more time to study the issue.

Don Felix Medina Tzuc and his son Gabriel, both active on the ejido council, were among the strong advocates of privatization. They argued that privatization will provide security of tenure and of property. This in turn will encourage serious farmers to do the long-term development that is now the only realistic way to make a living from farmland. Orchards, reforestation, beekeeping, irrigation, and livestock rearing are the primary developments to consider.

Most of the other voters were, like Don Felix, older men, serious farmers, and aware of the wider world and its trends. The nonvoters included almost all the women, and the younger and much older men, as well as nonfarmers and most of the small subsistence farmers.

Privatization has some obvious problems. The first is the practical question of how to divide up the land. The ejido needs to clear and definitively survey its borders. It has to find government surveyors who can lay out the plots. Also, it voted, in a hasty bit of mathematical challenge, to give every ejido family 46 hectares—not an easy matter when there are 330 families and only 14,330 ha to divide (15,180 ha would be needed). On top of this, the ejido plans to maintain a 1,700-ha forest reserve and (following the classic township model of the United States in the nineteenth century) some further lands for schools.

More serious, because less easily resolved, is the problem of fairness. Chunhuhub lands range from the best to the worst. The best lie in the mecanizada, the strip south of town along the main road, where irrigation and tractor use has opened up superb alluvial soils to intensive agriculture. The citrus corridors stretching south from there are also both highly fertile and already well developed. At the other extreme are bare-rock tablelands and the infertile, almost unworkable red clay of the southeastern corner of the ejido. These are doubly cursed in that they are largely inaccessible by current roads. Plans are afoot to divide land such that everyone has at least some access to both the good and the ill, but fairness will be virtually impossible to assure. The subsistence farmers have little capital to deploy in developing roads, wells, and other necessary infrastructure. The citrus orchardiers have a joint corporation set up to manage and hold their orchards, taking those lands out of the pool available for allocation. Plans suggest the possibility of dividing up the mecanizada separately from the rain-fed lands. Clearly, the future will bring much negotiating over this issue, and the ejido lands will not be divided until this is done.

Still further down the road is the problem that has surfaced in previous privatization cases the world around: families trying strategies that fail, forcing them into selling their lands and living in poverty and ruin. This problem is often compounded when such families are forced into selling land to rapacious outside interests that follow the "rape, ruin, and run" strategy. The ejido, and thus the Maya people as a whole, will likely lose more and more land to powerful outsiders who will often be less than well disposed toward local interests.

Many farmers are aware of these problems, but are convinced that the benefits of secure tenure, notably including ability to protect one's trees and animals, outweigh the risks. Behind all this lies a realization, not explicit and probably not consciously admitted by most, that subsistence maize farming is a fading trail. Population rises steadily, beyond the capability of subsistence agriculture to support. Droughts and violent storms are more frequent. This fits exactly the predictions of global warming models. Whether they ascribe the changes to global warming or the wrath of God, the Maya are facing the uncomfortable realization that rainfall-fed maize farming may grow less and less viable for the indefinite future. Indeed, a sharp drought was devastating to the 2004 maize crop, and, ominously, the asamblea began with formal applications for government relief by all the small farmers. In the outer fringes of Chunhuhub, where dwell the least affluent subsistence farmers, the dogs were skeletally thin, a sure sign of desperate want. Maya take care of their dogs, which become a gauge of family welfare.

With all this, the old codes of management are breaking down. Boys are shooting birds with slingshots—or, perhaps equally significant, are refraining from shooting them because of what they learn in school rather than because of what they learn from elders at home and in the field. Knowledge of medicinal plants is fading, knowledge of modern medicine rapidly growing. Ceremonies for the old gods become rarer and simpler. Spanish is increasingly the language of even the outer fringes of town.

Maya culture and agriculture flourish, and are in no danger of dying, but the infinite subtleties and details of managing the forest are rarely learned by today's children.

Another realization also lies unmentioned but is much more clearly known to all: education and urbanization bulk larger and larger in every family's future, every year. Most families have someone in town sending back remittances. Many ejidatarios live from shops and businesses and have given up serious farming. Others have purchased large private holdings outside the ejido and gone into intensive agriculture for the market.

Indeed, Chunhuhub is changing. A large new health center, with many more facilities, has replaced the tiny clinic. Don Salomon Xool, purveyor of *cochinita pibil* (pig baked in the earth oven) and other wonderful foods, has left his open-air temporary pitch and renovated an abandoned building downtown, turning it into a beautiful new taco restaurant.

Farther afield, the spectacular Classic Maya sites at Margueritas and Nueva Loria–Altamirano are finally being surveyed and excavated by professional archaeologists. Reconstruction would make them tourist destinations potentially equivalent to Coba and Tulum. They could dramatically change the fortunes of west-central Quintana Roo. Sadly gone will be the monkeys, king vultures, and other rare rain-forest birds and mammals that took shelter in these sites. However, the gains to knowledge, artistic heritage, and local economy will be great.

In Felipe Carrillo Puerto, two doors down from the huge old church built by the independent Cruzoob Maya as the center of their world, the Balam Nah ("House of the Jaguar") Computer Center now flourishes. It not only repairs and sells computers, but creates Web pages. One fears that the only jaguars housed in the area today are on those Web pages. On Balam Nah's doors are posters for the Sian Ka'an Biosphere Reserve, and advertisements for cultural tourism in the town of Señor, where many of the last rebel Maya held out till the end. The end, it appears, is to exhibit their traditions to the world—in some ways a victory, in some a humiliating loss.

Back in Chunhuhub, farmers still bicycle out to remote milpas, machetes in hand. Men still hunt deer and pacas in trackless forests. Women wear flowery *huipiles* and spend hours a day making tortillas. At the same time, more and more of the young are going to school for more and more years, and now many of them do not need to leave town to find jobs; Chunhuhub is becoming a town of teachers, businesspeople, health workers, mechanics, and modern market-oriented farmers.

And Don Felix's granddaughter Maybi is going to college to become an anthropologist.

Appendix A
The Secret World According to Doña Elsi Ramirez

DOÑA ELSI is my next-door neighbor in Chunhuhub. She is a widow; her husband was killed by a poisonous snake many years ago. As of 2001, Doña Elsi had been in Chunhuhub for forty years and was fifty-nine years old. She came from southern Yucatán state. Her aunt was a main source of lore.

Her mother died by falling from a high hammock. She broke a bone in her arm and it festered. One grandmother died from a hand injury—fell from a horse, it hit her, the wound festered. "Woman is like a chaya plant" (she quoted an old saying)—women's limbs break easily. She says that children used to be raised much more strictly, and the saying was that if you don't castigate your children, Jesus will castigate you. Boys and girls used to be shy with each other. (In fact, the tolerance shown children "these days" is more a difference of geography than of generation; Quintana Roo Maya are much more laid-back child raisers than are those of Elsi's home area.) A devil lived in a baalche' tree here and kept everyone awake. An old woman drove it away.

Rancho Tampac, formerly known as Paytoro because a bull's head appeared from the water in the well there, has a treasure: three pots of buried silver. One man sought it and disappeared. It is guarded by a huge snake. Xnibacal, another cattle ranch, has ancient sites. (One man asked me if I could find hidden treasure with my binoculars.) Spirit armadillos, goats, huge cows, chickens, and the *kuulkalk'iin*—a spirit with its head cut off—appear if you go after the Tampac treasure. With a proper oración you can get it, though. One man got treasure from a sinkhole here. In Iturbide a golden burro appeared. In Polyuc, gold chains and silver stuff was found; a man reburied it, paid for a mass for the dead who left it, and got it—if he'd just taken it, it would have disappeared—or worse!

A rolling ball of white cotton appeared in the church once, and a black humanoid being near it.

Dogs and opossums can be ghosts of people.

The Yumilk'aax is the Lord of the Forest—one person—known in Spanish as Juan de Monte.

Fortunately less often seen is the *chaap*—a giant, monkeylike being, who has lots of dogs. He is usually not dangerous. The *kulpach'* is similar. (These are the Quintana Roo version—also known as *sinsimito*—of the worldwide "wild man" or "bigfoot" figure.)

Xul (a town in far southern Yucatán—possibly named because it is, in every way, the end [*xul*] of the road) has many *aires*. She lived there and at night had to give corn gruel to the aluuxo'ob.

There are *way* witches in the form of goats, cats, and dogs. A Xtabay was seen walking from Tuc's corner northward. She hears the Yuntsiloob whistling around the place. These whistles cause disease if you're in their way. She has heard four whistle in a square around here. One should give the Yuntsiloob corn gruel at the milpa so that they will keep snakes from biting.

Aluxoob are winds that look like children. They are rare in Chunhuhub, but common in Xul. They steal ashes or food.

There is a *mal viento* that sounds like a young chicken seeking its mother. It causes headache and other diseases, and you must hold a *santiguar* ritual to clear it away.

The navel of Mani is the navel of the world. Others say Mérida's navel is the center. These navels are large navel-shaped rocks. There is a bottomless cave in Mani.

Cave water, called "virgin water," is magical. There is a magic well near Mani, the *saabaj ch'een*. If you drink a cocoyol shell of its water, you'll be saved when the world ends. An old woman had a vision of this. Another well in Mani connects to Merida by a kind of subway crossing. Only a brief walk and you're in Merida cathedral. A rope there was cut and blood came out of it: the umbilical cord of the world, or of something comparably important. Stone crosses come alive at night in Mani and so does a stone woman.

There is a live banana plant in the earth in Loltun Cave; a Virgin of Pilar was taken from there. (This probably refers to the fig that grows in the cave; somebody may have carved the Virgin out of wood from the tree.) *Ch'akxiix*—cave water dripping from stalactites into stone vessels made by hollowing out stalagmites—is extremely powerful. (Cf. *Diccionario Cordemex*, under *xix*. Many stalagmites in caves were broken off and carved into shallow bowls in ancient times by the Maya. Some have been broken off long enough to grow sizable new stalagmites in the bowls.)

Pantak'iintsul is a cave with resident spirits. Many birds go to drink and get cool water near the entrance.

A *lo'ojij* (probably from *loj*, and the same as, or cognate with, *lojil*, "redemption") is a person or saint who stays in a church and cures people.

Iturbide—a town in Yucatán—has many ruins, and also many Yuntsiloob, who scare people around sinkholes. A drunk was eaten there by the spirits because he offended a *saay* (leafcutter ant) hill that was associated with the feet of the Yuntsiloob.

Some people put rue in their eyes every Friday, thus getting the ability to send the Evil Eye. Their children all die. Some people can kill you in two to three hours by giving the Eye. Normally, the Eye is involuntarily given, usually by drunks, sometimes by pregnant women or sick people.

Some people are lazy due to mal viento—she has a cousin with this condition. This is not a case of *nervios* or just laziness, but a sickness of evil origin.

A jmeen buried chile, *ya'axjalalche'*, and some corn gruel at each corner of her place, and a chicken in the middle; so there are no aires here.

Elsi told the common folktale of the dove and the squirrel, in which the squirrel tricks the dove and eats her young, thus causing the dove's continual mourning. The melancholy-sounding call of the white-winged dove does indeed sound like *ku'uk tu tusen*, "the squirrel tricked me." In

Elsi's version the story ends with the squirrel getting her eggs—much more sensible than the usual version, in which the young are eaten. Squirrels are notorious egg robbers but are not outright carnivores.

Witches put frogs and snakes in your belly with food, also cause cancer that isn't really cancer, and can make trees fall on you, or the like.

An evil wind hitting your back or other areas of the body causes pains and diseases.

There are nine classes of herbs (nine is the Spanish magic number).

When her husband died of snakebite, he was far out in the forest. He had been warned of danger there. After that, his spirit bothered her. The soul of the dead goes to Heaven, but the aire of the body wanders around. If she fails to keep his grave clean, her house gets dirty for no apparent reason. He had had premonitions of his death.

THE FIRST TIME I used this medicine [an herbal cure] with my children was for congestion [from asthma]. My aunt came. She says for me not to give the kids other medicines. She sought her herb, she put it in a comal. It was *bakej ak'*. It was toasted in the comal. After that, I did like this [rubs hands together] and got the juice out. This I mixed in water—put some in—and put in nine drops of balsam [patent medicine], and gave a spoonful to my child. Then she afterward toasted another. I used my hands to get out the broth [juice], and bathed his stomach and lung area, and then bathed everything from his feet to his head with the medicine. I covered him and he slept. Then he had no more congestion. He got healthy with this medicine.

All my children were troubled with asthma. With this medicine I raised them. But Angelica, with this medicine she didn't get healthy—no. [Another] medicine, called *pujuk*, was boiled. And she took it warm. [In] a bowl. And it healed her.

Sometimes when a child is born, its navel sticks out, as big as this [gesture indicates about an inch]. As if it was pushed. So you seek the proper herbs. Epazote, oregano, cloves, cumin, and allspice and garlic. You mash them, with deer suet. You make a little tortilla of this, heat it, and put it on the navel. You do this for nine days and the navel heals.

Here is the medicine for vomiting: take an orange leaf, and one from orange bergamot mint, and spearmint. If you have balsam you add it in. Boil water and make the medicine [put the leaves quickly in the boiling water]. Cover it and take it warm.

For *k'alwiix* [difficulty in urination]: when a person can't urinate, has strangury, you take 'elemuuy [root] and mash it. After that you take root [probably a mistake for "fruit"] of the *kat'* [*Parmentiera aculeata*, a well-known diuretic] tree and mash it. Wash the dirt off and boil. And take nine hairs of an agouti. Put them in and boil. When it's cold, you take a small cupful. It cleans out the bladder of a person who can't urinate; one is healed with this.

For ojo ["evil eye," actually infant disorders such as diarrhea and infant jaundice]: clean a bottle and fill it with water. When your child begins to get close to dying, take pure water, put it in the bottle. When your child is near death, the water in the bottle gets yellow. The reflection is gone in the bottle in the light. [This seems to mean that the color of the disease, here jaundice of the newborn, goes off into the bottle.] Because there are times when the baby has green diarrhea, sometimes from the sun, sometimes from ojo. This cures it. [Apparently the bottle draws off the jaundice into the water.]

FELIX: This is knowledge that we had about sickness and remedies for curing the children. There were times when there was no money, and if you didn't know some remedies, you could cure nothing and your children would die.

Notes

PREFACE

1. The name is from *chun* (*chuun* in modern transcription), "trunk," and *huhub* (*jujuub*) "wild plum, *Spondias mombin*." This plum is not related to the temperate-zone plum, but it bears similar fruit. It is sometimes called the hog-plum in English, but this is a slander on an excellent fruit. Ralph Roys thought that *huhub* meant *Pinus caribaea* (pine; Roys 1976:246); pines do not even occur in most of Quintana Roo. Chunhuhub was named for the huge wild plums that used to grow there. Smaller ones still abound. The domestic *Spondias purpurea, abal* in Maya, also abounds, as an orchard tree.

2. There have been important short works on cultural ecology and agriculture by Ellen Kintz (1990), Carmen Morales Valderrama (1987), and Margarita Rosales González (1988—a historical study that provides the background for Morales's work). Following these, there appeared the more ambitious work of Betty Faust (1998), which adds a deep study of symbolism and religion to thorough research on current and historical environmental trends in interior Campeche. Yvan Breton and Delfin Quezada studied the fishing communities of the Yucatán coast (Quezada Dominguez 1995), while Alfredo César Dachary and Stella Maris Arnaiz Burne did the same for Quintana Roo (Cesar Dachary and Arnaiz Burne 1985). Sociologists Othon Baños Ramírez (1989, 1990) and Eric Villanueva Mukul (1985, 1990) reviewed political ecology. (See also the review of Maya studies in García Mora and Villalobos Salgado 1988, vol. 15.)

Farther afield, there are Carlos Inchaustegui's valuable studies of the Chontal Maya of Tabasco, and the research by Atran (1993; Atran et al. 1999) and Schwartz (1990) on the Itza, and Richard Wilk's study of the Kekchi of Belize (Wilk 1991).

Unpublished theses include, most notably, Hilaria Maas Colli's insightful account of child rearing and socialization (1983) and María Elena Peraza Lopez's study of foodways (1986). Juan Sosa's study of modern Maya cosmology (1985) and Ueli Hostettler's account of Yaxley in eastern Quintana Roo (Hostettler 1996; see also Hostettler 1993) also remain in thesis form. Valladolid, long neglected, has received attention (Gongora Biachi and Ramirez Carrillo 1993). Other themes include folk art (Terán 1994; Terán and Rasmussen 1981) and religion (Forrest 2004; Love and Peraza Castillo 1984; Love 1991; Faust 1998). Antonio Higuera Bonfil (e.g., 1997) and Luz del Carmen Vallarta Velez (e.g., 1997) have studied modern religious change. Local culture and tourism is also relevant (Daltabuit 1992; Daltabuit and Pi-Sunyer 1990; Daltabuit et al. 1988; Faust 1998; Castañeda 1996).

Scholars who are themselves Yucatec Maya are carrying out current research in all areas (e.g., Balam 1992; Arzápalo Marín 1987, 1995).

Maya speakers increased from 31,816 in 1990 to 37,212 in 1995 in Felipe Carrillo Puerto municipio (INEGI 2000c:13), out of 56,000 total population, and that number has been increasing since.

Quintana Roo has 873,804 people as of 2000 (INEGI 2000b:9). Of those over the age of five, 174,700 speak Yucatec (INEGI 2001:379). This figure—22.9 percent of those in that age group—is not as high as Yucatán's 37.8 percent, which is actually the highest percentage of any state, and a total of 587,300 people. The total Yucatec speaker population reported in 2000 is 776,824 (INEGI 2001:383). Though nothing from Yucatán has achieved the fame of Rigoberta Menchu's autobiography from Guatemala, the Yucatec have taken full advantage of their chances to speak (Burns 1983, 1990; Sullivan 1989; see also Acuña 1993 and Barrera Vásquez, 1980, 1981, 1989). This literature has found a historian in Francesc Ligorred Perramon, a Catalan Mayanist, in his book *U Mayathanoob Ti Dzib/Las voces de la escritura* (i.e., *Maya Writings* [Maya title], *Voices from What Is Written* [Spanish title], 1997). Carlos Montemayor (1995) and Manuel Gutiérrez-Estévez (1998) have paid attention to surviving traditional Maya literature. David Bolles published several very valuable folktales from Kom Chen, many of them told by his wife, a local individual (Bolles 1985; Bolles and Kim de Bolles 1973; Kim de Bolles 1973). Noteworthy among them is a brief story to the effect that God is getting tired of the world and throws his cigar butts at it—these are what we call shooting stars; the Virgin Mary prevents them landing on us and burning us up; but she too is getting "fed up," and one day . . . (Bolles 1985:40–41). Mayanists will immediately think of the aged cigar-smoking high god of the pre-Columbian Maya, "God K," who is obviously somewhere behind this story.

Margaret Park Redfield (1935) may deserve the distinction of being the pioneer of serious studies by "outsiders" of Maya oral literature. Moreover, the Redfields worked with the linguist Manuel Andrade, who collected a vast store of folktales (Andrade and Maas Colli 1990).

Still more impressive was William Hanks' study of everyday Maya language use, *Referential Practice* (1990).

No textbook of Maya is widely available. (A brief nineteenth-century effort [Zavala 1974/1896] is still sold, but it is of little use.) Bilingual schools have really excellent texts, notably the *Maaya T'aan* series (Secretaría de Educación Publica 1996), but these cannot be found outside of the schools. Briefly available was Hilaria Maas Colli's introductory text (Maas Colli 1995). This was published (by the University of Yucatán) in a tiny press run and never republished. When I went to get a copy, about a year after publication, I was told by staff at the press outlet in Mérida that they had never heard of such a book. Virtually no libraries found out about the book in time to obtain copies. There remain, in manuscript or locally published form, Robert Blair and Refugio Vermont Solis's *Spoken Yucatec Maya* (1965) and David Bolles's *A Grammar of Yucatecan Maya* (1973). Bricker et al. (1998) provide a superb but very local dictionary, as well as publications on grammar and usage.

Much of the best work on the Yucatec Maya languishes in unpublished theses and reports. Even the published work is often hard to find. Terán and Rasmussen's book (1994) on milpa agriculture, a work not only scholarly but also well written and beautifully illustrated with Dr. Rasmussen's photographs, was issued by the State of Yucatán in a press run of only five hundred copies. Within a year the book was impossible to find on the open market, and only through the good offices of Drs. Terán and Rasmussen was I able to obtain a copy. The same fate—a tiny press run by a state government, and instant disappearance—befell what is probably the greatest single book on the south Mexican countryside, Miguel Alvarez del Toro's *Así Era Chiapas* (1985). State presses tend to give many of their copies to dignitaries and officials, and these copies often are lost.

CHAPTER 1

1. This work draws on the theories of cultural ecology (Steward 1955), especially those now often called "political ecology" (see, e.g., Sheridan 1995; Stonich 1993). The latter term is not without challenge; Andrew Vayda and Bradley Walters (1999) have claimed it privileges politics too much, leading people to ignore biology and other matters. The present book is based on a broad framework of the sort Julian Steward (1955) envisioned.

2. Communist-led countries have a worse record than capitalist ones; they have invoked even more ecologically destructive agricultural processes, often without even increasing production in the process (see, e.g., Smil 1984, 1993). The "free market" (to say nothing of that other idealization of the free, Murray Bookchin's ecological anarchism; Bookchin 1982, 1988) might well work—if it could really exist anywhere. However, the "free market" is an ideal type, not a possible reality.

CHAPTER 3

1. A typical old-time milpa—Teodomiro Tun Xool's in 1991, to be exact—uses 20 kg of maize seed, 1 ½ kg beans, and 1 kg squash seeds per ha. He cut 4 ha: 3 *caña* (*sak'al*, cropped the year before), 1 in forest. Don Teodomiro also has a parcela of 2 ha with chaya, citrus, banana, coconut, mamey, mango, manioc, tamarind, achiote, sugarcane, papaya, caimito, avocado, sweet potatoes, and *xa'an* palmettos (left from the wild). This is near milpas and parcelas of his father and other relatives. The main Tun complex is southwest of Rancho Tampac, while the main Xool fields are on the flat "mecanizada" land southeast of it.

2. Government pamphlets advise a range of insecticides, including Diazinon, Parathion, Malathion, Afidrin, Morestan, Chlordane (brand names)—most of which are banned in the United States, and some in Mexico! Pests mentioned include aphids *Aphis spiraecola* and other spp., white plantlouse *Prontaspis ganonensis*, blackfly *Aleurocanthus woglumi*, mite *Tetranychus sexmaculatus*, and of course the indefatigable leafcutters *Atta* spp. Only the last is a problem in Chunhuhub. The others are strongly suspected to be problems only where overuse of insecticide has knocked out their natural enemies. Diseases of citrus trees include *Phythophthora citrophthora*, a root disease that is serious only in wet, infected soils; *Micosphaerella citri*; and the newly arrived tristeza virus. The former two are not problematic. Tristeza, however, is getting to be a genuinely frightening problem. It is fatal and incurable. It is vectored by aphids. Government pamphlets advise heavy use of dangerous pesticides, which would eliminate all possibility of biological control. In general, a citrus orchard that followed the government schedules would be in serious trouble in short order. Fortunately, the Maya are both too sensible and too poor to invest so heavily in pesticides, and predators therefore continue to give adequate control of the aphids—giving time for the Maya to plant tristeza-resistant varieties. Still, the disease is a problem. (It is more a problem in Pich than in Chunhuhub [Betty Faust, personal communication 2001], presumably because Pich is more receptive to modernization and chemicals.)

3. Fortunately, cattle numbers are leveling off in the area. Reported cattle population in the municipio declined from 6,300 to 5,768 from 1993 to 1998 and 6,050 in 1999 (INEGI 2000a:215). I am certain this is a large undercount, but, indeed, cattle have not boomed in the area. There are, as of 1999, 135,820 cattle reportd for Quintana Roo; most of these are in the central south (INEGI 2000a:215). One can contrast Yucatán state, with 624,488 (INEGI 2000b:299). Vast cattle deserts occupy most of Yucatán's agricultural land—489,514 out of 787,801 ha (INEGI 2000b:271; it is true that the biggest single chunk of this is around Tizimin, where the savannahs are very old and may be at least partly natural).

4. From the other side of the Peninsula, we have the sad news that palm-leaf hats are going out (Chuc Uc 1999). These provided an incentive to preserve palmettos. Sabal palms have been managed very well in Quintana Roo; they are preserved when milpa is cut, and so tend to increase over time, especially since they are fire resistant. They have not been so fortunate in Yucatán state (Caballero 1991) or in Cuba, where they being overharvested seriously, to the detriment of birds that nest in them (Wechsler 1988).

CHAPTER 4

1. Mahogany and cedro produce about 8 cubic meters of wood per tree of minimum legal size (55 cm dbh); the biggest trees can yield up to 20 cubic meters in Quintana Roo (and much bigger farther south and southeast). A cubic meter of barked log, straight from the forest, is figured at 225 board feet. A cubic meter is valued at around 1,600–2,000 pesos. Sawn into lumber, it is worth about 12 pesos, which yields 6.5 pesos per board foot in net income to the mill.

Ordinary tropical woods (*corrientes tropicales*) are worth about half that. However, some, notably granadillo and ciricote, are even more valuable than mahogany and cedro.

CHAPTER 5

1. To the point that phenomenology becomes a justification for insane Nazism, as in the cases of Martin Heidegger and Paul de Man. The ethical implications of this have been traced out by Emmanuel Levinas (1961), who comes to a startling and profound set of conclusions about the extreme and fundamental importance of interpersonal relations. Levinas' moral view is strikingly close to that of the Maya.

2. The Quintana Roo Maya view is, however, very much like that of many old-time farmers of the United States and the British Isles—including my own farming ancestors, from whom I learned a worldview that I now find I share with the Maya. These farmers did not have the traditional animistic beliefs of the Maya; their views were based on Protestant concepts of stewardship and responsibility. These Protestant attitudes have become the views of many contemporary Maya, who have found Protestantism conformable with their traditional valuation of animals and plants. The utter destructiveness and ruthlessness toward nature that now characterizes Anglo-American farming is of some antiquity, but was by no means universal until the rise of heavily subsidized giant-corporation farming, absentee landlordism, and other social ills.

3. They are not depleting groundwater. Runoff sinks immediately into the porous limestone and flows away underground. The pumps thus tap a flow—a great underground river.

CHAPTER 6

1. Anonymous children's rhyme, published by Gabriel Zaid (1995:52) and reprinted here by his kind permission; my translation is somewhat free, to keep the rhyme.

CHAPTER 7

1. Notable was the late Dra. Yolanda Ramos Bravo, who died tragically young from a heart attack in 1997, and who resided part time in Chunhuhub in 1996. She also covered Tepich, Xpichil, Tihosuco, and various smaller communities in her practice. She said that Chunhuhub and Xpichil (pop. ca. 1,400) are in good shape, but Tepich (pop. ca. 3,000)—in a poor, densely populated, dry area—is far worse off. Some people there could not walk because of chronic malnutrition, and stunting is common.

References

Abramovitz, Janet N. 2001. *Unnatural Disasters*. Washington, D.C.: Worldwatch Institute.

Acosta Bustillos, Elena, J. Salvador Flores Guido, and Arturo Gómez-Pompa. 1998. *Uso y manejo de plantas forrajeras para cria de animales dentro del solar en una comunidad Maya en Yucatán*. Mérida, Mexico: Universidad Autónoma de Yucatán.

Acuña, Rene, ed. 1993. *Bocabulario de Maya Than*. Mexico City: Universidad Autónoma de México.

Alcocer Puerto, Elias Miguel. 2001. "Manejo sustentable de recursos naturales y culturales por parte de una comunidad Maya de Yucatán: El caso de Yaxunah." Licenciado thesis, social anthropology, Universidad Autónoma de Yucatán, Mérida.

Alcorn, Janis. 1984. *Huastec Mayan Ethnobotany*. Austin: University of Texas Press.

Almanza Alcalde, Horacio. 2000. "Percepciones locales de la naturaleza en el área de protección de flora y fauna 'Yum Balam' en Quintana Roo." Licenciada thesis, social anthropology, Universidad Autónoma de Yucatán, Mérida.

Altieri, M. 1980. "Diversification of Corn Agroecosystems as a Means of Regulating Fall Armyworm Populations." *Florida Entomologist* 63:450–56.

Alvard, Michael. 1995. "Intraspecific Prey Choice by Amazonian Hunters." *Current Anthropology* 36:789–818.

Alvard, Michael, John G. Robinson, Kent H. Redford, and Hillard Kaplan. 1997. "The Sustainability of Subsistence Hunting in the Neotropics." *Conservation Biology* 11:977–82.

Alvarez del Toro, Miguel. 1985. *¡Así era Chiapas!*. Tuxtla Gutiérrez, Mexico: Universidad Autónoma de Chiapas.

———. 1991. *Los mamíferos de Chiapas.* 2d ed. Tuxtla Gutierrez, Mexico: Instituto Chiapaneco de Cultura.

Ancona y Rivera, Guadalupe, Suemi McLiberty Martínez, Lorena Careaga Viliesid, Adriana de Castro, Elsy Rodríguez Carrillo, Arturo Bayona, Francisco Javier Ayuso Zetina, and Ramón Iván Suárez Caamal. 1995. *Dinámica social de Quintana Roo.* Mexico City: Ediciones Pedagógicas.

Anderson, Barbara, E. N. Anderson, Tracy Franklin, Aurora Dzib Xihum de Cen. 2004. "Pathways of Decision Making among Yucatan Maya Traditional Birth Attendants." *Journal of Midwifery and Women's Health* 49, 4:312-19.

Anderson, E. N. 1988. *The Food of China.* New Haven: Yale University Press.

———. 1991. "Resource Management in a Maya Village." Paper presented at the Maya Conference, University of California, Riverside.

———. 1992a. "Folk Forestry in a Maya Town." Paper presented at the Annual Conference of the Southwestern Anthropological Association, San Diego, Calif.

———. 1992b. "Can Ejidos Work? Forest Management in a Maya Town." Paper presented at the Annual Conference of the International Association for the Study of Common Property, Washington, D.C.

———. 1992c. "Animal Names in a Maya Town." Paper presented at the Biennial Conference of the International Society of Ethnobiology, Columbia, Mo.

———. 1993. "Gardens in Tropical America and Tropical Asia." *Biotica, Nueva Epoca,* 1:81–102.

———. 1996a. *Ecologies of the Heart.* New York: Oxford University Press.

———. 1996b. "Gardens of Chunhuhub." In *Los Mayas de Quintana Roo: Investigaciones antropologicas recientes,* ed. Ueli Hostetler, 64–76. Bern, Switzerland: Universitat Bern, Institut für Ethnologie, Arbeitsblatter #14.

———. 2000. "Maya Knowledge and 'Science Wars.'" *Journal of Ethnobiology* 20:129–58.

———. 2001. "The Morality of Ethnobiology." Paper presented at the Annual Meeting of the Society of Ethnobiology, Durango, Colo.

———. 2003. *Those Who Bring the Flowers: Ethnobotany of Western Quintana Roo.* Chetumal, Mexico: ECOSUR.

Anderson, E. N., and Marja L. Anderson. 1973. *Mountains and Water: The Cultural Ecology of South Coastal China.* Taipei: Orient Cultural Service.

———. 1978. *Fishing in Troubled Waters.* Taipei: Orient Cultural Service.

Anderson, E. N., and Felix Medina Tzuc. 2004. *Animals and the Maya in Southeast Mexico.* Tucson: University of Arizona Press.

Anderson, Myra, and E. N. Anderson. 1996. "Food for the Rain Gods." Paper presented at the Annual Meeting of the American Anthropological Association, San Francisco.

Andrade, Manuel J., and Hilaria Maas Colli, eds. 1990. *Cuentos Mayas Yucatecos.* Mérida, Mexico: Universidad Autónoma de Yucatán.

Andrews Heath de Zapata, Dorothy. 1979. *El Libro del Judio o Medicina Domestica.* Mérida, Mexico: Dorothy Andrews Heath de Zapata.

Ankli, Anita. 2000. "Yucatec Mayan Medicinal Plants: Ethnobotany, Biological Evaluation, and Phytochemical Study of *Crossoptetalum gaumeri.*" Doctor of Natural Sciences thesis, Swiss Federal Institute of Technology, Zurich, Switzerland.

Ankli, Anita, Otto Sticher, and Michael Heinrich. 1999a. "Medical Ethnobotany of the Yucatec Maya: Healers' Consensus as a Quantitative Criterion." *Economic Botany* 53:144–60.

———. 1999b. "Yucatec Maya Medicinal Plants Versus Nonmedicinal Plants: Indigenous Characterization and Selection." *Human Ecology* 27:557–80.

Arias, Luis M., D. E. Williams, L. Latoumerie, E. Sauri, G. Balam, J. L. Chavez, and A. Ramírez. 2003. "Maize and Man in the Maya of Southern Mexico: Comparing Local and Scientific Classifications of Yucatán Landraces." Paper presented at the Annual Meeting of the Society for Economic Botany, Tucson, Ariz.

Arvigo, Rosita, and Michael Balick. 1993. *Rainforest Remedies: One Hundred Healing Herbs of Belize.* Twin Lakes, Wis.: Lotus Press.

Arzápalo Marín, Ramon. 1995. *Calepino de Motul.* Mexico City: Universidad Autónoma de México.

Arzápalo Marín, Ramon, ed. 1987. *Ritual de los Bacabes.* Mexico City: Universidad Autónoma de México.

Arzápalo Marín, Ramon, and Ruth Gubler, eds. 1997. *Persistencia cultural entre los Mayas frente al cambio y la modernidad.* Mérida, Mexico: Universidad Autónoma de Yucatán.

Ascher, William. 1999. *Why Do Governments Waste Natural Resources?* Baltimore, Md.: Johns Hopkins University Press.

Atran, Scott. 1993. "Itzá Maya Tropical Agro-Forestry." *Current Anthropology* 34:633–700.

———. 1999. "Itzaj Maya Folkbiological Taxonomy: Cognitive Universals and Cultural Particulars." In *Folkbiology*, ed. Douglas Medin and Scott Atran, 119–204. Cambridge, Mass.: MIT Press.

Atran, Scott, Douglas Medin, Elizabeth Lynch, Valentina Vapnarsky, Edilberto Ucan Ek', and Paulo Sousa. 2001. "Folkbiology Doesn't Come from Folkpsychology: Evidence from Yukatek Maya in Cross-Cultural Perspective." *Journal of Cognition and Culture* 1:5–42.

Atran, Scott, Douglas Medin, Norbert Ross, Elizabeth Lynch, John Coley, Edilberto Ucan Ek', and Valentina Vapnarsky. 1999. "Folkecology and Commons Management in the Maya Lowlands." *Proceedings of the National Academy of Sciences* 96:7598–603.

Avendaño y Loyola, Fr. Andres. 1987. *Relation of Two Trips to Peten.* Culver City, Calif.: Labyrinthos.

Balam, Gilberto. 1992. *Cosmogonia y uso actual de las plantas medicinales de Yucatán.* Mérida, Mexico: Universidad Autónoma de Yucatán.

Baños Ramírez, Othón. 1989. *Yucatán: Ejidos sin campesinos.* Mérida, Mexico: Universidad Autónoma de Yucatán.

———. 1990. *Sociedad, estructura agraria y estado en Yucatán.* Mérida, Mexico: Universidad Autónoma de Yucatán.

Barlett, Peggy. 1982. *Agricultural Choice and Change.* New Brunswick, N.J.: Rutgers University Press.

Barlett, Peggy, ed. 1980. *Agricultural Decision Making.* New York: Academic Press.

Barrera, Alfredo, and Alfredo Barrera Vásquez. 1983. *El Libro del Judio.* Xalapa, Mexico: Instituto Nacional de Investigaciones sobre Recursos Bioticos.

Barrera Marín, Alfredo, Alfredo Barrera Vásquez, and Rosa María Franco. 1976. *Nomenclatura Etnobotánica Maya.* Mérida, Mexico: Instituto Nacional de Antropología e Historia.

Barrera Vásquez, Alfredo. 1980. "Sobre la unidad de habitacion tradicional campesina y el manejo de recursos bioticos en el area Maya Yucatánense, Pt. I." *Biotica* 5, no. 3:115–29.

———. 1981. *Estudios Lingüísticos.* Mérida, Mexico: Fondo Editorial de Yucatán.

———. 1989. El libro de los libros de Chilam Balam. Mérida, Mexico: Producción Editorial Dante.

Barrera Vásquez, Alfredo, A. Gómez-Pompa, and C. Vazquez-Yanes. 1977. "El manejo de las selvas por los Mayas: sus implicaciones silvicolas y agricolas." *Biotica* 2, no. 2:47–61.

Bartolome, Miguel. 1988. *La dinamica social de los mayas de Yucatán: Pasado y presente de la situación colonial.* Mexico City: Instituto Nacional Indigenista.

Benedict, F. G., and M. Steggerda. 1936. *The Food of the Present-day Maya Indians of Yucatán.* Washington, D.C.: Carnegie Institution, Publication 456, Contribution 18.

Benitez, Fernando. 1986 [1956]. *Ki: El drama de un pueblo y una planta.* Mexico City: Fondo de Cultura Económica.

Bennett, John. 1982. *Of Time and the Enterprise.* Minneapolis: University of Minnesota Press.

Berkes, Fikret. 1999. *Sacred Ecology*. Philadelphia: Taylor and Francis.

Berlin, Brent, and Elois Ann Berlin. 1996. *Medical Ethnobiology of the Highland Maya of Chiapas, Mexico: The Gastrointestinal Diseases*. Princeton, N.J.: Princeton University Press.

Berlin, Brent, Dennis Breedlove, and Peter Raven. 1974. *Principles of Tzeltal Plant Classification*. New York: Academic Press.

Blair, Robert, and Refugio Vermont Solis. 1965. "Spoken (Yucatec) Maya." Mimeograph, University of Chicago, Chicago.

Bolles, David. 1985. "World Creation Stories from Kom Cheen, Yucatán." In *Mexicon*. N.p.

Bolles, David, with Alejandra Kim de Bolles. 1973. *A Grammar of Yucatecan Maya*. Komchhen, Mexico: David Bolles. (Includes "Mayan Folk Tales," by Alejandra Kim de Bolles.)

Bookchin, Murray. 1982. *The Ecology of Freedom*. Palo Alto, Calif.: Cheshire Books.

———. 1988. "Social Ecology Versus Deep Ecology: A Challenge for the Ecology Movement." *Socialist Review* 88:11–29.

Botkin, Daniel. 1990. *Discordant Harmonies: A New Ecology for the Twenty-First Century*. New York: Oxford University Press.

Bovard, James. 1989. *The Farm Fiasco*. San Francisco: ICS Press.

Bracamonte y Sosa, Pedro. 1993. *Amos y sirvientes: Las haciendas de Yucatán, 1789–1860*. Mérida, Mexico: Universidad Autónoma de Yucatán.

———. 1994. *La memoria enclaustrada: Historia indígena de Yucatán, 1750–1915*. Mérida, Mexico: Universidad Autónoma de Yucatán.

Brannan, Jeffery T., and Gilbert M. Joseph, eds. 1991. *Land, Labor, and Capital in Modern Yucatán: Essays in Regional History and Political Economy*. Tuscaloosa: University of Alabama Press.

Bray, David Barton, Laticia Merino-Pérez, Patricia Negreros-Castillo, Gerardo Segura-Wiarnholtz, Juan Manuel Tores-Rojo, and Henricus F. M. Vester. 2003. "Mexico's Community-Managed Forests as a Global Model for Sustainable Landscapes." *Conservation Biology* 17:672–77.

Breedlove, Dennis, and Robert Laughlin. 1993. *The Flowering of Man*. Washington, D.C.: Smithsonian Institution Press.

Bricker, Victoria. 1981. *The Indian Christ, the Indian King*. Austin: University of Texas Press.

Bricker, Victoria, Eleuterio Po'ot Yah, and Ofelia Dzul de Po'ot. 1998. *A Dictionary of the Maya Language as Spoken in Hocabá, Yucatán*. Salt Lake City: University of Utah Press.

Brown, Denise. 1984. "Estudio Sociocultural de la Comunidad Chontal de Tucta, Tabasco." Ms.

———. 1988. Catálogo de Comunidades Mayas en Tres Zonas de la Península de Yucatán. Ms.

———. 1993. "Yucatec Maya Settling, Settlement and Spatiality." Ph.D. dissertation, Dept. of Anthropology, University of California, Riverside.

Brown, Lester. 1995. *Who Will Feed China? Wake-up Call for a Small Planet.* New York: W. W. Norton.

Brown, Lester, Michael Renner, and Christopher Flavin. 1998. *Vital Signs.* New York: W. W. Norton.

Brown, Michael. 1998. "Can Culture Be Copyrighted?" *Current Anthropology* 39:193–222.

———. 2003. *Who Owns Native Culture?* Cambridge, Mass.: Harvard University Press.

Burger, Joanna, Elinor Ostrom, Richard Norgaard, David Policancsky, and Bernard D. Goldstein, eds. 2001. *Protecting the Commons.* Washington, D.C.: Island Press.

Burns, Allan. 1983. *An Epoch of Miracles.* Austin: University of Texas Press.

———. 1990. "There Will Come a Day: Present Day Mayan Thoughts about Ecological Collapse." Manuscript circulated by Institute of Maya Studies.

Caballero, Javier. 1991. "Procesos de cambio en la interacción hombre-planta. El caso del uso y manejo de la palma del guano entre los mayas de Yucatán, México." Paper presented at the Second International Congress of Ethnobotany, Kunming, China.

———. 1997. "Origin and Evolutionary Tendencies of the Traditional Homegardens of the Tropical Lowlands of Mexico." Ms.

Callicott, J. Baird. 1994. *Earth's Insights.* Berkeley: University of California Press.

Cancian, Frank. 1979. *The Innovator's Situation: Upper Middle-Class Conservatism in Agricultural Communities.* Palo Alto, Calif.: Stanford University Press.

Canopy. 1998. "From Forest to Fret: The Making of a Smartwood Guitar" (from "Smart Sounds: Music for the Planet II Concert Program"). *Canopy,* March-April, pp. 1–2. New York: Rainforest Alliance.

Carter, William E. 1969. *New Lands and Old Traditions: Kekchi Cultivators in the Guatemalan Lowlands.* Gainesville: Center for Latin American Studies, University of Florida.

Castañeda, Quetzil. 1996. *In the Museum of Maya Culture.* Minneapolis: University of Minnesota Press.

Castro, María Cristina, Gabriel Macías, Antonio Higuera, and Luz del Carmen Vallarta. 1986. *Quintana Roo: Procesos políticos y democracia.* Mexico City: CIESAS.

Caulfield, Richard. 1997. *Greenlanders, Whales, and Whaling*. Hanover, N.H.: Dartmouth College and University Press of New England.

César Dachary, Alfredo, and Stella Maris Arnaiz Burne. 1985. *El Caribe Mexicano: Hombres e Historias*. Mexico City: Museo Nacional de Culturas Populares.

————. 1992. *Bitácora de un viaje a la justicia: Crónicas de una huelga olvidada*. Chetumal, Mexico: Centro de Investigaciones sobre Quintana Roo.

Chase, Arlen, and Diana Chase. 1999. "Scale and Intensity in Classic Period Maya Agriculture: Terracing and Settlement at the 'Garden City' of Caracol, Belize." *Culture and Agriculture* 20:60–77.

Chavelas Polito, Javier. 1981. "El *Pinus caribaea* Morelet, en el Estado de Quintana Roo, México" (leaflet). Mexico City: Instituto Nacional de Investigaciones Forestales, SARH.

Chayanov, A. V. 1966. *On the Theory of Peasant Economy*. Ed. and trans. Daniel Thorner, Basile Kerblay, and R. E. F. Smith. Homewood, Ill.: American Economic Association.

Chuc Uc, Cessia Esther. 1999. "El tejido des sombreros de jipi en Becal, Campeche: Auge y decadencia (1930–1997)." Licenciatura thesis, Antropología Social, Universidad Autónoma de Yucatán.

Clendinnen, Inga. 1987. *Ambivalent Conquests*. Cambridge: Cambridge University Press.

Climo, Jacob. 1977. Collective Farming in Northern and Southern Yucatán, Mexico: Ecological and Administrative Determinants of Success and Failure. *American Ethnologist* 4:191–205.

Coe, Michael. 1992. *Breaking the Maya Code*. New Haven, Conn.: Yale University Press.

Colding, Johan, and Carl Folke. 2001. "Social Taboos: 'Invisible' Systems of Local Resource Management and Biological Conservation." *Ecological Applications* 11:584–600.

Conference on Common Property Resource Management. 1986. *Proceedings*. Washington, D.C.: National Academy of Sciences Press.

Conklin, Harold. 1957. *Hanunoo Agriculture*. Rome, Italy: FAO.

Contreras Sánchez, Alicia del C. 1990. *Historia de una tintorea olvidada: El proceso de explotacion y circulacion del palo de tinte 1750–1807*. Mérida, Mexico: Universidad Autónoma de Yucatán.

Cortes, Hernan. 1986. *Letters from Mexico*. Ed. and trans. Anthony Pagden. New Haven, Conn.: Yale University Press.

Crooks, Deborah L. 1997. "Biocultural Factors in School Achievement for Mopan Children in Belize." *American Anthropologist* 99:586–601.

Crosby, Alfred. 1972. *The Columbian Exchange: Biological and Cultural Consequences of 1492*. Westport, Conn.: Greenwood Press.

Culbert, T. Patrick, ed. 1973. *The Classic Maya Collapse*. Albuquerque: University of New Mexico Press.

Curtis, Jason H., David A. Hodell, and Mark Brenner. 1996. "Climate Variability on the Yucatán Peninsula (Mexico) during the Past 3500 Years, and Implications for Maya Cultural Evolution." *Quaternary Research* 46:37–47.

Daltabuit, Magalí. 1992. "Mujeres mayas: Trabajo, nutrición y fecundidad." Mexico City: Instituto de Investigaciones Antropológicas.

Daltabuit, Magalí, Alicia Rios Torres, and Fraterna Perez Plaja. 1988. *Coba: Estrategias adaptativas de tres familias Mayas*. Mexico City: Universidad Autónoma de México.

Daltabuit, Magalí, and Oriol Pi-Sunyer. 1990. "Tourism Development in Quintana Roo, Mexico." *Cultural Survival Quarterly* 14, no. 1:9–13.

Darch, J. P., ed. 1983. *Drained Field Agriculture in Central and South America*. Oxford: BAR.

del Amo, Silvia, and Raul Aguilar Lojero. 1987. "The Tecallis: A Traditional Water Management System." Ms.

Demarest, Arthur, Matt O'Mansky, Claudia Wolley, Dirk Van Tuerenhout, Takeshi Inomata, Joel Palka, and Hector Escobedo. 1997. "Classic Maya Defensive Systems and Warfare in the Petexbatun Region: Archaeological Evidence and Interpretations." *Ancient Mesoamerica* 8:229–54.

Denevan, William. 1976. *The Native Population of the Americas in 1492*. Madison: University of Wisconsin Press.

———. 2001. *Cultivated Landscapes of Native Amazonia and the Andes*. New York: Oxford University Press.

Díaz-Bolio, José. 1974a. *La Chaya*. Mérida, Mexico: José Diaz-Bolio.

———. 1974b. *El Libro de los Guisos de Chaya*. Mérida, Mexico: José Diaz-Bolio.

Dichter, Thomas. 2003. *Despite Good Intentions: Why Development Assistance to the Third World Has Failed*. Amherst: University of Massachusetts Press.

Dobyns, Henry. 1983. *Their Number Became Thinned*. Nashville: University of Tennessee Press.

Doolittle, William. 2000. *Cultivated Landscapes of Native North America*. New York: Oxford University Press.

Dumond, Don. 1998. *The Machete and the Cross*. Lincoln: University of Nebraska Press.

Durkheim, Emile. 1995 [1912]. *The Elementary Forms of Religious Life*. Trans. Karen Fields. New York: Free Press.

Durrenberger, E. Paul, ed. 1984. *Chayanov, Peasants, and Economic Anthropology.* New York: Academic Press.

Edwards, C. R. 1986. "The Human Impact on the Forest in Quintana Roo, Mexico." *Journal of Forest History* 30, no. 3:120–27.

Eliade, Mircea. 1964. *Shamanism: Archaic Techniques of Ecstasy.* New York: Pantheon.

Ellen, Roy. 1994. *The Cultural Relations of Classification.* Cambridge: Cambridge University Press.

Elster, Jon. 1983. *Explaining Technical Change.* Cambridge: Cambridge University Press.

Emch, Michael. 2003. "The Human Ecology of Mayan Cacao Farming in Belize." *Human Ecology* 31:111–32.

Erickson, Clark. 1987. "Prehistoric Intensive Agriculture and Indigenous Communities in the Lake Titicaca Basin of Peru." Paper presented at the Conference on Traditional Wet Field Agriculture in the American and Asian Tropics, University of California, Riverside.

Escobar, Arturo. 1995. *Encountering Development: The Making and Unmaking of the Third World.* Princeton, N.J.: Princeton University Press.

Escobar, Arturo. 1999. "After Nature: Steps to an Antiessentialist Political Ecology." *Current Anthropology* 40:1–30.

Farrington, Ian S., ed. 1985. *Prehistoric Intensive Agriculture in the Tropics.* Oxford: BAR.

Farris, Nancy. 1984. *Maya Society under Colonial Rule.* Princeton, N.J.: Princeton University Press.

Faust, Betty. 1989. "Tractors and Collective Strategies of the Campeche Maya." Paper presented at the Annual Meeting of the American Anthropological Association, Washington, D.C.

———. 1991. "Maya Culture and Maya Participation in the International Ecotourism and Resource Conservation Project." In *Ecotourism and Resource Conservation.*, ed. Jon A. Kusler, 178–221. Mérida, Mexico: Pronatura.

———. 1998. *Maya Rural Development and the Plumed Serpent.* Westport, Conn.: Bergin and Garvey.

———. 2001. "Maya Environmental Successes and Failures in the Yucatán Peninsula." *Environmental Science and Policy* 4:153–69.

Faust, Betty Bernice, and Richard Bilsborrow. Forthcoming. "Maya Culture, Population, and the Environment in the Yucatán Peninsula." In *Population-Environment Interactions on the Yucatán Peninsula,* ed. Wolfgang Lutz, Leonel Prieto, and Warren Sanderson. Laxenburg, Austria: International Institute for Applied Systems Analysis.

Faust, Betty, E. N. Anderson, and John Frazier, eds. 2004. *Rights, Resources, Culture, and Conservation in Maya Communities of the Yucatán*. Westport, Conn.: Greenwood Press.

Faust, Betty, and John Sinton. 1991. "Let's Dynamite the Salt Factory." In *Ecotourism and Resource Conservation*, ed. Jon A. Kusler, 602–24. Mérida, Mexico: Pronatura.

Fedick, Scott. 1996a. "Introduction: New Perspectives on Ancient Maya Agriculture and Resource Use." In *The Managed Mosaic*, ed. Scott Fedick, 1–14. Salt Lake City: University of Utah Press.

———. 1996b. "An Interpretive Kaleidoscope: Alternative Perspectives on Ancient Agricultural Landscapes of the Maya Lowlands." In *The Managed Mosaic*, ed. Scott Fedick, 107–31. Salt Lake City: University of Utah Press.

Fedick, Scott, ed. 1996. *The Managed Mosaic*. Salt Lake City: University of Utah Press.

Feldman, Lawrence H., ed. and trans. 2000. *Lost Shores, Forgotten Peoples: Spanish Explorations of the South East Maya Lowlands*. Durham, N.C.: Duke University Press.

Flachsenberg, Henning, and Hugo A. Galletti. 1998. "Forest Management in Quintana Roo, Mexico." In *Timber, Tourists, and Temples: Conservation and Development in the Maya Forest of Belize, Guatemala, and Mexico*, ed. Richard Primack, David Bray, Hugo A. Galletti, and Ismael Ponciano, 47–60. Washington, D.C.: Island Press.

Flannery, Kent V. 1972. The Cultural Evolution of Civilizations. *Annual Review of Ecology and Systematics* 3:399–426.

Flannery, Kent V., ed. 1982. *Maya Subsistence: Studies in Memory of Dennis E. Puleston*. New York: Academic Press.

Flores Guido, J. Salvador. 1983. "Significado de los haltunes (sartenejas) en la cultura Maya." *Biotica* 8, no. 3:259–79.

———. 1984. *Algunas formas de caza y pesca usadas en Mesoamerica*. Xalapa, Mexico: Instituto Nacional de Investigaciones sobre Recursos Bioticos.

———. 1987. *Uso de los recursos vegetales en la peninsula de Yucatán: Pasado, presente y futuro*. Xalapa, Mexico: Instituto Nacional de Investigaciones sobre Recursos Bioticos.

Flores Guido, J. Salvador, and Ileana Espejel Carvajal. 1994. *Tipos de Vegetación de la Peninsula de Yucatán*. Mérida, Mexico: Universidad Autónoma de Yucatán and Sostenibilidad Maya.

Flores Guido, J. Salvador, and Jesús Kantún Balam. 1997. "Importance of Plants in the *Ch'a' Chaak* Maya Ritual in the Peninsula of Yucatán." *Journal of Ethnobiology* 17:97–108.

Flores Guido, J. Salvador, and Edilberto Ucan Ek. 1983. "Nombres usados por los Mayas para designar a la vegetación." Xalapa, Mexico: Instituto Nacional de Investigaciones sobre Recursos Bioticos.

Folan, William J., Laraine A. Fletcher, and Ellen R. Kintz. 1979. "Fruit, Fiber, Bark, and Resin: Social Organization of a Maya Urban Center." *Science* 204:697–701.

Forrest, David W. 2004. "The Landscape of Santa Rita Komchhen: The Role of Metaphor in the Ecological History of a Yucatecan Rural Estate." In *Rights, Resources, Culture, and Conservation in Maya Communities of the Yucatán*, ed. Betty Faust, E. N. Anderson, and John Frazier, 175-90. Westport, Conn.: Greenwood Press.

Foster, George. 1964. "Treasure Tales and the Image of the Static Economy in a Mexican Peasant Community." *Journal of American Folklore* 77:394.

———. 1965. "Peasant Society and the Image of Limited Good." *American Anthropologist* 67:293–315.

———. 1994. *Hippocrates' Latin American Legacy: Humoral Medicine in the New World*. Langhorne, Pa.: Gordon and Breach.

Frank, Robert. 1988. *Passions within Reason*. Cambridge, Mass.: Harvard University Press.

Franquemont, Christine. 1988. "The Mnemonics of Chinchero Botany: How Children Learn and Adults Remember the Natural World." Paper presented at the Annual Conference of the Society for Ethnobiology, Mexico City.

Frazier, Jack. 1997. "The 'Tectetan' Syndrome." Paper presented at the Annual Meeting of the American Anthropological Association, Washington, D.C.

———. 2004. "The 'Yucatán' Syndrome: Its Relevance to Biological Conservation and Anthropological Activities." In *Rights, Resources, Culture, and Conservation in Maya Communities of the Yucatán*, ed. Betty Faust, E. N. Anderson, and John Frazier, 225-54. Westport, Conn.: Greenwood Press.

Freese, Curtis H., ed. 1997. *Harvesting Wild Species: Implications for Biodiversity Conservation*. Baltimore, Md.: Johns Hopkins University Press.

Galletti, Hugo A. 1998. "The Maya Forest of Quintana Roo: Thirteen Years of Conservation and Community Development." In *Timber, Tourists, and Temples: Conservation and Development in the Maya Forest of Belize, Guatemala, and Mexico.*, ed. Richard Primack, David Bray, Hugo A. Galletti, and Ismael Ponciano, 33–46. Washington, D.C.: Island Press.

García, Hernán, Antonio Sierra, and Gilberto Balám. 1996. *Medicina Maya tradicional: Confrontación con el sistema conceptual Chino*. Mexico City: Educación, Cultura y Ecología A.C.

García Mora, Carlos, and Martín Villalobos Salgado, eds. 1988. *La antropología en México*. Vol. 15: *La antropología en el sur de México*. Mexico City: Instituto Nacional de Antropología e Historia.

García Sancho, Guillermo Macías. 1985. *La Leyenda del Alux*. Chetumal, Mexico: Consejo Editorial del Instituto Quintanarroense de la Cultura.

de la Garza, Mercedes, Ana Luisa Izquierdo, María del Carmen Leon, and Tolita Figueroa, eds. 1983. *Relaciones Historico-geograficas de la Gobernacion de Yucatán*. Vol. 1: *Mérida, Valladolid y Tabasco*. Mexico City: Universidad Autónoma de México.

Gates, Marilyn. 1992. *In Default: Peasants, the Debt Crisis, and the Agrarian Challenge in Mexico*. Boulder, Colo.: Westview.

Geertz, Clifford. 1963. *Agricultural Involution: The Process of Ecological Change in Indonesia*. Berkeley: University of California Press.

Gill, Richardson. 2000. *The Great Maya Droughts*. Albuquerque: University of New Mexico Press.

Gladwin, Christina. 1977. "A Model of Farmers' Decisions to Adopt the Recommendations of Plan Puebla." Ph.D. dissertation, Food Research Institute, Stanford University.

Gleick, Peter. 2000. *The World's Water: The Biennial Report on Freshwater Resources, 2000–2001*. Washington, D.C.: Island Press.

Godoy, Ricardo. 1999. "The Difference a Year Makes." Letter published in *Science* 285:1850.

Goldkind, Victor. 1965. "Social Stratification in the Peasant Community: Chan Kom Revisited." *American Anthropologist* 67:863–84.

———. 1966. "Class Conflict and Cacique in Chan Kom." *Southwestern Journal of Anthropology* 22, no. 4:325–45.

Gómez-Pompa, Arturo. 1985. *Los Recursos Bioticos de México: Reflexiones*. Mexico City: Instituto Nacional de Investigaciones sobre Recursos Bioticos.

———. 1987. "On Maya Silviculture." *Mexican Studies/Estudios Mexicanos* 3, no. 1:1–17.

Gómez-Pompa, Arturo, M. F. Allen, S. L. Fedick, and J. J. Jiménez-Osornio, eds. 2003. *The Lowland Maya Area: Three Millennia at the Human-Wildland Interface*. New York: Haworth Press.

Gómez-Pompa, Arturo, José Salvador Flores, and Victoria Sosa. 1987. "The 'Pet Kot': A Man-made Tropical Forest of the Maya." *Interciencia* 12, no. 1:10–15.

Gómez-Pompa, Arturo, and Andrea Kaus. 1990. "Traditional Management of Tropical Forests in Mexico." In *Alternatives to Deforestation*, ed. Anthony Anderson, 45–64. New York: Columbia University Press.

Gómez-Pompa, Arturo, C. Vasquez-Yanes, and S. Guevara. 1972. "The Tropical Rain Forest, a Nonrenewable Resouce." *Science* 177:762–65.

Gongora Biachi, Renan A., and Luis A. Ramirez Carrillo, eds. 1993. *Valladolid: Una ciudad, una region y una historia*. Mérida, Mexico: Universidad Autónoma de Yucatán.

Gonzalez, Roberto. 2001. *Zapotec Science*. Austin: University of Texas Press.

Goodstein, Eban. 1999. *The Trade-off Myth: Fact and Fiction about Jobs and the Environment*. Washington, D.C.: Island Press.

Green, Donald, and Ian Shapiro. 1994. *Pathologies of Rational Choice Theory*. New Haven, Conn.: Yale University Press.

Greenberg, Joseph. 1987. *Language in the Americas*. Palo Alto, Calif.: Stanford University Press.

Gutiérrez-Estévez, Manuel. 1998. "Plurality of Perspectives and Subjects in the Literary Genres of the Yucatec Maya." *American Anthropologist* 100:309–25.

Habermas, Jurgen. 1984 [1981]. *The Theory of Communicative Action*. Trans. Thomas McCarthy. Boston: Beacon Press.

Hancock, Graham. 1991. *Lords of Poverty: The Power, Prestige, and Corruption of The International Aid Business*. London: Mandarin (Reed International).

Hanks, William. 1990. *Referential Practice*. Chicago: University of Chicago Press.

Hardin, Garrett. 1968. "The Tragedy of the Commons." *Science* 162:1243–48.

———. 1991. "The Tragedy of the *Unmanaged* Commons." In *Commons without Tragedy: Protecting the Environment from Overpopulation, a New Approach*, ed. Robert V. Andelson, 162–85. Lanham, Md.: Rowman and Littlefield.

Harris, Marvin. 1966. The Cultural Ecology of India's Sacred Cattle. *Current Anthropology* 7:51–66.

———. 1968. *The Rise of Anthropological Theory*. New York: Thomas Crowell.

Harrison, P., and B. L. Turner. 1978. *Pre-Hispanic Maya Agriculture*. Albuquerque: University of New Mexico Press.

Haug, Gerald H., Detlef Günther, Larry C. Peterson, Daniel M. Sigman, Konrad A. Hughen, and Beat Aeschlimann. 2003. "Climate and the Collapse of Maya Civilization." *Science* 299:1731–35.

Hayami, Yujiro, and Vernon Ruttan. 1985. *Agricultural Development*. 2d ed. Baltimore, Md.: Johns Hopkins University Press.

Headland, Thomas. 1997. "Revisionism in Ecological Anthropology." *Current Anthropology* 38:605–30.

Hellmuth, N. 1977. "Cholti-Lacandon (Chiapas) and Peten-Ytza Agriculture, Settlement Pattern, and Population." In *Social Process in Maya Prehistory: Studies in Honour of Sir Eric Thompson*, ed. Norman Hammond, 421–45. London: Academic Press.

Herrera Castro, Natividad. 1994. "Los huertos familiares Mayas en el oriente de Yucatán." Mérida, Mexico: Universidad Autónoma de Yucatán y Sostenibilidad Maya.

Hervik, Peter. 1999. *Mayan People within and beyond Boundaries*. Amsterdam: Harwood Academic Publishers.

Higuera Bonfil, Antonio. 1997. "Los testigos de Jehová en el Caribe mexicano." Paper presented at the 49th International Congress of Americanists, Quito, Ecuador.

Hodell, David, Mark Brenner, Jason Curtis, and Thomas Guilderson. 2001. "Solar Forcing of Drought Frequency in the Maya Lowlands." *Science* 292:1367–70.

Hofling, Charles Andrew, with Félix Fernando Tesucún. 1997. *Itzaj Maya-Spanish-English Dictionary*. Salt Lake City: University of Utah Press.

Hostettler, Ueli. 1992. "New Inequalities: Socioeconomic Change among Central Quintana Roo Maya." Paper presented at the Annual Meeting of the American Anthropological Association, San Francisco.

———. 1993. "Unidad doméstica y estratificación socioeconómica: El caso de los mayas cruzoob en el centro del Estado de Quintana Roo, México." Ms.

———. 1996. "Milpa Agriculture and Socioeconomic Diversification: Socioeconomic Change in a Maya Peasant Society of Central Quintana Roo, 1900–1995." Ph.D. dissertation, University of Berne, Switzerland.

———. 2002. "Labor Regime and Social Justice: Consequences of Economic and Social Stratification among Maya Peasants in Central Quintana Roo, Mexico." *Anthropos* 97:107–16.

Houston, Stephen. 1997. "The Shifting Now." *American Anthropologist* 99:291–305.

Hughes, J. Donald. 1982. *American Indian Ecology*. El Paso: Texas Western University Press.

Hultkrantz, Ake. 1967. *The Religions of the American Indians*. Berkeley: University of California Press.

Hunn, Eugene. 1977. *Tzeltal Folk Zoology*. New York: Academic Press.

———. 1982. "Mobility as a Factor Limiting Resource Use in the Columbia Plateau of North America." In *Resource Managers*, ed. Nancy Williams and Eugene Hunn, 17–44. Boulder, Colo.: Westview.

Instituto Nacional de Estadistica Geografia e Informatica (INEGI). 1987. *Anuario estadistico del Estado de Quintana Roo, 1986*. Mexico City: INEGI.

———. 1991. *Resultados preliminares, XI censo general de poblacion y vivienda, 1990*. Mexico: INEGI.

———. 1994. *Quintana Roo: Resultados definitivos, VII censo agricola-ganadero*. Mexico: INEGI.

———. 1995. *Anuario estadistico del Estado de Quintana Roo*. Mexico: INEGI.

———. 2000a. *Anuario estadistico: Quintana Roo*. Edición 2000. Mexico: INEGI.

———. 2000b. *Anuario estadistico: Yucatán*. Edición 2000. Mexico: INEGI.

———. 2000c. *Cuaderno estadistico municipal. Felipe Carrillo Puerto, Estado de Quintana Roo*. Edición 2000. Mexico: INEGI.

———. 2000d. *Estadisticas del medio ambiente*. 2 vols. Mexico: INEGI.

———. 2001. *Mujeres y Hombres en México*. Mexico: INEGI.

Islebe, Gerald A., Henry Hooghiemstra, Mark Brenner, Jason H. Curtis, and David A. Hodell. 1996. "A Holocene Vegetation History from Lowland Guatemala." *The Holocene* 6:265–71.

Jerome, Norge, Randy Kandel, and Gretel Pelto. 1980. *Nutritional Anthropology*. Nw York: Redgrave.

Jiménez Osornio, Juan. 1987. "Chinampas in the Valley of Mexico Today." Paper presented at the Traditional Wetfield Agriculture in the American and Asian Tropics Conference, University of California, Riverside.

Jones, Grant D. 1989. *Maya Resistance to Spanish Rule*. Albuquerque: University of New Mexico Press.

———. 1991. *El Manuscrito Can Ek*. Mérida, Mexico: Instituto Nacional de Antropología e Historia and National Geographic Society.

de Jong, Harriet J. 1999. "The Land of Corn and Honey." Ph.D. dissertation, University of Utrecht, Netherlands.

Jorgensen, Jeffrey. 1994. "La caceria de subsistencia practicada por la gente Maya en Quintana Roo." In *Madera, chicle, caza y milpa: Contribuciones al manejo integral de las selvas de Quintana Roo, Mexico*, ed. Laura K. Snook and Amanda Barrera de Jorgensen, n.p. Chetumal, Mexico: ECOSUR.

———. 1998. "The Impact of Hunting on Wildlife in the Maya Forest of Mexico." In *Timber, Tourists, and Temples: Conservation and Development in the Maya Forest of Belize, Guatemala, and Mexico*, ed. Richard Primack, David Bray, Hugo A. Galletti, and Ismael Ponciano, 179–94. Washington, D.C.: Island Press.

Juarez, Ana. 2002. "Ecological Degradation, Global Tourism, and Inequality: Maya Interpretations of the Changing Environment in Quintana Roo, Mexico." *Human Organization* 61:113–24.

Kaus, Andrea. 1992. "Common Ground: Ranchers and Researchers in the Mapimí Biosphere Reserve." Ph.D. dissertation, Dept. of Anthropology, University of California, Riverside.

Kay, Charles. 2001. "Afterword: False Gods, Ecological Myths, and Biological Reality." In *Wilderness and Political Ecology*, ed. Charles Kay and Randy Simmons, 238–61. Salt Lake City: University of Utah Press.

Kay, Charles, and Randy Simmons, eds. 2001. *Wilderness and Political Ecology*. Salt Lake City: University of Utah Press.

Kearney, Michael. 1984. *World View*. Novato, Calif.: Chandler and Sharp.

———. 1996. *Reconceptualizing the Peasantry*. Boulder, Colo.: Westview.

Keohane, Robert, Michael McGinnis, and Elinor Ostrom. 1992. *Proceedings of a Conference on Linking Local and Global Commons*. Cambridge, Mass., and Bloomington, Ind.: Harvard University and Indiana University.

Kiernan, Michael J., and Curtis H. Freese. 1997. "Mexico's Plan Piloto Forestal: The Search for Balance between Socioeconomic and Ecological Sustainability." In *Harvesting Wild Species: Implications for Biodiversity Conservation*, ed. Curtis H. Freese. Baltimore, Md.: Johns Hopkins University Press.

Killion, Thomas, ed. 1992. *Gardens of Prehistory: The Archaeology of Settlement Agriculture in Greater Mesoamerica*. Tuscaloosa: University of Alabama Press.

Kim de Bolles, Alejandra. 1973. *Mayan Folk Tales*. (Bound with Bolles and Kim de Bolles 1973.) Komchhen, Mexico: David Bolles.

Kimber, Clarissa. 1966. "The Dooryard Gardens of Martinique." *Yearbook of the Association of Pacific Coast Geographers* 28:97–118.

Kingsolver, Barbara. 2002. *Small Wonder: Essays by Barbara Kingsolver*. New York: HarperCollins.

Kinsley, Michael. 1995. *Ecology and Religion*. Englewood Cliffs, N.J.: Prentice-Hall.

Kintz, Ellen. 1990. *Life under the Tropical Canopy: Tradition and Change among the Yucatec Maya*. Fort Worth, Tex.: Holt, Rinehart and Winston.

Kirch, Patrick V. 1994. *The Wet and the Dry: Irrigation and Agricultural Intensification in Polynesia*. Chicago: University of Chicago Press.

———. 1997. "Microcosmic Histories." *American Anthropologist* 99:31–42.

Klee, Gary A., ed. 1980. *World Systems of Traditional Resource Management*. New York: V. H. Winston and Sons.

Konrad, Herman W. 1991. "Capitalism on the Tropical-Forest Frontier: Quintana Roo, 1880s to 1930." In *Land, Labor, and Capital in Modern Yucatán: Essays in Regional History and Political Economy*, ed. Jeffrey T. Brannan and Gilbert M. Joseph, 143–71. Tuscaloosa: University of Alabama Press.

Krech, Shepard. 1999. *The Ecological Indian: Myth and History*. New York: W. W. Norton.

Kunen, Julie L., T. Patrick Culbert, Vilma Fialko, Brian R. McKee, and Liwy Grazioso. 2000. "*Bajo* Communities: A Case Study from the Central Peten." *Culture and Agriculture* 22:15–31.

Laird, Sarah, ed. 2002. *Biodiversity and Traditional Knowledge: Equitable Partnerships in Practice*. London: Earthscan.

Landa, Diego de. 1937. *Yucatán Before and After the Conquest*. Trans. William Gates. Baltimore, Md.: Maya Society.

Lansing, Stephen. 1991. *Priests and Programmers*. Princeton, N.J.: Princeton University Press.

Las Casas, Bartolome de. 1992 [1542]. *A Short Account of the Destruction of the Indies*. Ed. and trans. Anthony Pagden. London: Penguin.

Lawrence, T. E. 1935. *Seven Pillars of Wisdom*. New York: Doubleday, Doran and Co.

Lazos Chavera, Elena. 1992. "From Corn to Oranges: Transformation of Traditional Management of Agricultural Systems in Oxkutzcab, Yucatán." Paper presented at the Annual Meeting of the International Society of Ethnobiology, Columbia, Mo.

Lenkersdorf, Carlos. 1996. *Los hombres verdaderos: Voces y testimonios tojolabales*. Mexico City: Siglo Veintiuno.

Lentz, David L., ed. 2000. *Imperfect Balance: Landscape Transformations in the Precolumbian Americas*. New York: Columbia University Press.

Levinas, Emmanuel. 1961. *Totality and Infinity*. Pittsburgh, Pa.: Duquesne University Press.

Ligorred Perramon, Francesc. 1997. *U Mayathanoob ti dzib/Las voces de la escritura*. Mérida, Mexico: Universidad Autónoma de Yucatán.

Linares, Olga. 1976. "Garden Hunting in the American Tropics." *Human Ecology* 4:331–50.

Llanes Pasos, Eleuterio. 1993. *Cuentos de cazadores*. Chetumal, Mexico: Government of Quintana Roo.

Lopez, Kevin Lee. 1992. "Returning to Fields." *American Indian Culture and Research Journal* 16:165–74.

López Austin, Alfredo. 1988. *The Human Body*. Salt Lake City: University of Utah Press.

Love, Bruce. 1989. "Yucatec Sacred Breads through Time." In *Word and Image in Maya Culture*, William Hanks and Don Rice, 336–50. Salt Lake City: University of Utah Press.

———. 1991. "Ancient Guardians of the Modern Milpa." Paper presented at the Conference on Ancient Maya Agriculture and Biological Resource Management, University of California, Riverside.

Love, Bruce, and Eduardo Peraza Castillo. 1984. "*Wahil kol*: A Yucatec Maya Agricultural Ceremony." *Estudios de Cultura Maya* 15:251–301.

Low, Bobbi S. 1993. "Behavioral Ecology of Conservation in Traditional Societies." Paper presented at the Annual Meeting of the American Anthropological Association, Washington, D.C.

Maas Colli, Hilaria. 1983. "Transmicion cultural: Chemax, Yucatán: Un enfoque etnografico." Thesis, Antropologia Social, University Autonoma de Yucatán, Mérida.

———. 1994. *Curso de lengua Maya para investigadores: Nivel principiantes.* Mérida, Mexico: Universidad Autónoma de Yucatán.

MacIntyre, Alasdair. 1988. *Whose Justice? Which Rationality?* Notre Dame, Ind.: Notre Dame University Press.

Mallory, Walter. 1926. *China, Land of Famine.* New York: American Geographic Society.

Mandeville, Bernard. 1924 [1732]. *The Fable of the Bees.* Oxford: Oxford University Press.

March M., Ignacio J. 1987. "Los Lacandones de México y su relacion con los mamiferos silvestres: un estudio etnozoologico." *Biotica* 12:43–56.

Marks, Robert. 1998. *Tigers, Rice, Silk, and Silt.* Cambridge: Cambridge University Press.

Marten, Gerald, ed. 1986. *Traditional Agriculture in Southeast Asia.* Honolulu: University of Hawaii Press.

Martin, Calvin. 1978. *Keepers of the Game.* Berkeley: University of California Press.

Marx, Karl. 1909 [1867]. *Capital,* vol. 1. Trans. G. Unterman. Chicago: C. H. Kerr.

Mata, Gerardo. 1987. "Introduccion a la etnomicologia Maya de Yucatán: El conocimiento de los hongos en Pixoy, Valladolid." *Revista mexicana de Micologia* 3:175–87.

McBryde, Felix. 1947. *Cultural and Historical Geography of Southwest Guatemala.* Publication 4, Institute of Social Anthropology. Washington, D.C.: Smithsonian Institution.

McCay, Bonnie, and James Acheson, eds. 1987. *The Question of the Commons.* Tucson, Ariz.: University of Arizona Press.

McCullough, John. 1973. "Human Ecology, Heat Adaptation, and Belief Systems: The Hot-Cold Syndrome of Yucatán." *Journal of Anthropological Research* 29:32–36.

McEvoy, Arthur. 1986. *The Fisherman's Problem.* Berkeley: University of California Press.

McGee, R. Jon. 1989. *Life, Ritual and Religion among the Lacandon Maya.* Belmont, Calif.: Wadsworth.

McJunkin, David. 1991. "Logwood: An Inquiry into the Historical Biogeography of *Haematoxylon campechianum* L. and Related Dyewoods of the Neotropics." Ph.D. dissertation, Department of Geography, University of California, Los Angeles.

Medin, Douglas, and Scott Atran, eds. 1999. *Folkbiology.* Cambridge, Mass.: MIT Press.

Meilleur, Brien. 1985. *Gens de Montagne: Plantes et Saisons: Termignon en Vanoise.* Le monde alpin et rhodanien #1. Grenoble, France: Centre Alpin et Rhodanien d'Ethnologie.

Melville, Elinor G. 1994. *A Plague of Sheep.* Cambridge: Cambridge University Press.

Mendieta, Rosa M., and Silvia del Amo. 1981. *Plantas Medicinales del Estado de Yucatán.* Mexico City: Compañia Editorial Continental S.A.

Messer, Ellen. 1978. *Zapotec Plant Knowledge: Classification, Uses, and Communication about Plants in Mitla, Oaxaca, Mexico.* Memoirs of the Museum of Anthropology, 10, part 2. Ann Arbor: University of Michigan.

Milton, Kay. 2002. *Loving Nature.* London: Routledge.

Mizrahi, Aliza, Francisco Xuluc Tolosa, Isabel Sohn, and Juan Jiménez-Osornio. 1996. "Conocimiento tradicional y aprovechamiento de las selvas de Sahcabá." *Desarollo Agroforestal y Comunidad Campesina* 26:46–51.

Moerman, Daniel. 1998. *Native American Ethnobotany.* Portland, Ore.: Timber Press.

Montemayor, Carlos. 1995. *Arte y composición en los rezos sacerdotales Mayas.* Mérida, Mexico: Universidad Autónoma de Yucatán.

Morales Valderrama, Carmen. 1987. *Ocupacion y sobrevivencia campesina en la zona citricola de Yucatán.* Mérida, Mexico: Instituto Nacional de Antropologia e Historia.

Morgan, David L. 1988. *Focus Groups.* Beverly Hills, Calif.: Sage Publications.

Morrell, Mike. 1985. *The Gitksan and Wet'suwet'en Fishery in the Skeena River System.* Hazelton, B.C.: Gitksan-Wet'suwet'en Tribal Council.

Morton, Julia. 1981. *Atlas of Medicinal Plants of Latin America.* Philadelphia: Charles C. Thomas.

Mossbrucker, Gudrun. 1997. "Lo que se sabe de la historia." Opiniones de la gente de Kantunil Kin, Quintana Roo, Mexico. Paper presented at the 49th International Congress of Americanists, Quito, Ecuador.

Murdoch, W. W. 1975. "Diversity, Stability, Complexity, and Pest Control." *Journal of Applied Ecology* 12:745–807.

Murphy, Julia. 1990. "Indigenous Forest Use and Development in the 'Maya Zone' of Quintana Roo, Mexico." Master's thesis, Environmental Science, York University, Ontario, Canada.

Nabhan, Gary P., Amadeo M. Rea, Karen L. Reichhardt, Eric Mellink, and Charles F. Hutchinson. 1982. "Papago Influences on Habitat and Biotic Diversity: Quitovac Oasis Ethnoecology." *Journal of Ethnobiology* 2:124–43.

Naess, Arne. 1986. "The Deep Ecological Movement: Some Philosophical Aspects." *Philosophical Inquiry* 8:1–20.

Nations, J. D., and R. B. Nigh. 1978. "Cattle, Cash, Food, and Forest: The Destruction of the American Tropics and the Lacandon Maya Alternative." *Culture and Agriculture* 6.

———. 1980. "The Evolutionary Potential of Lacandon Maya Sustained Yield Tropical Forest Agriculture." *Journal of Anthropological Research* 36:1–30.

Nelson, Richard. 1983. *Hunters of the Northern Forest.* Chicago: University of Chicago Press.

Netting, Robert. 1993. *Smallholders.* Palo Alto, Calif.: Stanford University Press.

North, Douglass. 1991. *Institutions, Institutional Change, and Economic Performance.* Cambridge: Cambridge University Press.

Oldenburg, Roy. 1989. *The Great Good Place.* New York: Paragon House.

Oldfield, Margery L., and Janis Alcorn. 1987. "Conservation of Traditional Agroecosystems." *BioScience* 37, no. 3:199–208.

Olson, Mancur. 1965. *The Logic of Collective Action.* Cambridge, Mass.: Harvard University Press.

Ortega Canto, Judith, Jolly Hoil Santos, and Angel Lendechy Grajales. 1996. *Leishmaniasis en milperos de Campeche (una aproximación médico-antropológica).* Mérida, Mexico: Universidad Autónoma de Yucatán.

Ortiz de Montellano, Bernard. 1990. *Aztec Medicine, Health, and Nutrition.* New Brunswick, N.J.: Rutgers University Press.

Ostrom, Elinor. 1990. *Governing the Commons.* Cambridge: Cambridge University Press.

Pagden, Anthony. 1982. *The Fall of Natural Man.* Cambridge: Cambridge University Press.

Painter, Michael, and William Durham, eds. 1995. *The Social Causes of Environmental Destruction in Latin America.* Ann Arbor: University of Michigan Press.

Parezo, Nancy, ed. 1993. *Hidden Scholars.* Albuquerque: University of New Mexico Press.

Patch, Robert. 1993. *Maya and Spaniard in Yucatán, 1648–1812.* Palo Alto, Calif.: Stanford University Press.

Peet, R. K. 1974. "The Measurement of Species Diversity." *Annual Review of Ecology and Systematics* 5:285–307.

Peraza López, María Elena. 1986. "Patrones alimenticios en Ichmul, Yucatán: Sus determinantes socioeconomicas, ecologicas y culturales." Licenciada thesis, Universidad Autónoma de Yucatán.

Peters, Charles M., Silvia E. Purata, Michael Chibnik, Berry J. Brosi, Ana M. López, and Myrna Ambrosio. 2003. "The Life and Times of *Bursera glabrifolia* (H.B.K.) Engl. in Mexico: A Parable for Ethnobotany." *Economic Botany* 57:431–41.

Pilcher, Jeffrey M. 1998. *Que vivan los tamales!: Food and the Making of Mexican Identity*. Albuquerque: University of New Mexico Press.

Pinkerton, Evelyn. 1989. *Cooperative Management of Local Fisheries*. Vancouver: University of British Columbia Press.

Pinkerton, Evelyn, and Martin Weinstein. 1995. *Fisheries That Work: Sustainability Through Community-Based Management*. Vancouver, B.C.: David Suzuki Foundation.

Piperno, Dolores, and Deborah L. Pearsall. 1998. *Origins of Agriculture in the Neotropics*. San Diego: Academic Press.

Pi-Sunyer, Oriol, and R. Brooke Thomas. 1997. "Tourism, Environmentalism, and Cultural Survival in Quintana Roo." In *Life and Death Matters*, ed. Barbara Rose Johnston, 187–212. Walnut Creek, Calif.: AltaMira.

Plotkin, Mark. 1993. *Tales of a Shaman's Apprentice*. New York: Penguin.

Pohl, Mary D., Kevin O. Pope, John G. Jones, John S. Jacob, Dolores R. Piperno, Susan deFrance, David L. Lentz, John A. Gifford, Marie E. Danforth, and J. Kathryn Josserand. 1996. "Early Agriculture in the Maya Lowlands." *Latin American Antiquity* 7:355–72.

Pope, Kevin, and Bruce Dahlin. 1989. "Ancient Maya Wetland Agriculture: New Insights from Ecological and Remote Sensing Research." *Journal of Field Archaeology* 16:87–106.

Prescott, William H. 1902 [1843]. *Mexico and the Life of the Conqueror Fernando Cortes*. New York: P. F. Collier and Son.

Press, Irwin. 1975. *Tradition and Adaptation: Life in a Modern Yucatán Maya Village*. Westport, Conn.: Greenwood Press.

Primack, Richard B., David Bray, Hugo A. Galletti, and Ismael Ponciano, eds. 1998. *Timber, Tourists, and Temples: Conservation and Development in the Maya Forest of Belize, Guatemala, and Mexico*. Washington, D.C.: Island Press.

Puleston, Dennis. 1978. "Terracing, Raised Fields, and Tree Cropping in the Maya Lowlands: A New Perspective on the Geography of Power." In *Pre-Hispanic Maya Agriculture*, ed. Peter Harrison and B. L. Turner, 225–46. Albuquerque: University of New Mexico Press.

Pulido Salas, María Teresa, and Lidia Serralta Peraza. 1993. *Lista anotada de las plantas medicinales de uso actual en el Estado de Quintana Roo, México*. Chetumal, Mexico: Centro de Investigaciones sobre Quintana Roo.

Quezada Dominguez, Delfin. 1995. *Papel y transformación de las unidades de producción pesquera ejidales en el sector halieútico, Yucatán, México*. Mérida, Mexico: Universidad Autónoma de Yucatán.

Ramirez Barajas, Pablo Jesus, and Nuria Torrescano Valle. 2000. *Uso y manejo de los recursos bioticos en la comunidad Maya de Petcacab, Quintana Roo*. Mexico City: Universidad Autónoma de México.

Randall, Robert. 1977. "Change and Variation in Samal Fishing: Making Plans to 'Make a Living' in the Southern Philippines." Ph.D. dissertation, University of California, Berkeley.

Rappaport, Roy A. 1971. "The Sacred in Human Evolution." *Annual Review of Ecology and Systematics* 2:23–44.

———. 1984. *Pigs for the Ancestors*. 2d ed. New Haven, Conn.: Yale University Press.

———. 1999. *Ritual and Religion in the Making of Humanity*. Cambridge: Cambridge University Press.

Re Cruz, Alicia. 1996. *The Two Milpas of Chan Kom*. Albany: State University of New York Press.

Redfield, Margaret Park. 1935. *The Folk Literature of a Yucatecan Town*. Contributions to American Archaeology #13. Washington, D.C.: Carnegie Institution.

Redfield, Robert. 1930. *Tepoztlan: Life in a Mexican Village*. Chicago: University of Chicago Press.

———. 1941. *The Folk Culture of Yucatán*. Chicago: University of Chicago Press.

———. 1950. *A Village that Chose Progress*. Chicago: University of Chicago Press.

Redfield, Robert, and Margaret Park Redfield. 1940. *Disease and Its Treatment in Dzitas, Yucatán*. Contributions to American Anthropology and History #32. Washington, D.C.: Carnegie Institution.

Redfield, Robert, and Alfonso Villa Rojas. 1934. *Chan Kom: A Maya Village*. Chicago: University of Chicago Press.

Redford, Kent H. 1990. "The Ecologically Noble Savage." *Cultural Survival Quarterly* 15:1, 46–48.

———. 1996. "Getting to Conservation." In *Traditional Peoples and Biodiversity Conservation in Large Tropical Landscapes*, ed. Kent H. Redford and Jane A. Mansour, 251–65. Arlington, Va.: The Nature Conservancy.

Redford, Kent H., and Jane A. Mansour, eds. 1996. *Traditional Peoples and Biodiversity Conservation in Large Tropical Landscapes*. Arlington, Va.: The Nature Conservancy.

Redman, Charles. 1999. *Human Impacts on Ancient Environments*. Tucson: University of Arizona Press.

Reed, Nelson. 1964. *The Caste War of Yucatán*. Palo Alto, Calif.: Stanford University Press.

Reichel-Dolmatoff, Gerardo. 1967. *Amazonian Cosmos*. Chicago: University of Chicago Press.

Remmers, G. G. A., and H. de Koeijer. 1992. "The T'olche', a Maya System of Communally Managed Forest Belts: The Causes and Consequences of Its Disappearance." *Agroforestry Systems* 18:149–77.

Restall, Matthew. 1997. *The Maya World*. Palo Alto, Calif.: Stanford University Press.

———. 1998. *Maya Conquistador*. Boston: Beacon Press.

Rice, Richard E., Raymond Gullison, and John Reid. 1997. "Can Sustainibile Management Save Tropical Forests?" *Scientific American*, April 1997, 44–49.

Rico Gray, Victor, A. Gómez-Pompa, and Castulo Chan. 1985. "Las selvas manejadas por los Mayas de Yohaltun, Campeche, Mexico." *Biotica* 10, no. 4:321–27.

Rico Gray, Victor, and Monica Palacios Rios. 1992. "Use and Management of the Deciduous Forest by the Maya of Central Yucatán, Mexico." Paper presented at the Biennial Meeting of the International Society of Ethnobotany, Columbia, Mo.

Ridington, Robin. 1981. "Technology, World View, and Adaptive Strategy in a Northern Hunting Society." *Canadian Review of Sociology and Anthropology* 19, no. 4:469–81.

Ridley, Matt. 1996. *The Origins of Virtue*. New York: Viking.

Roberts, Leslie. 1988. "Hard Choices Ahead on Biodiversity." *Science* 241:1759–61.

Rojas Rabiela, Teresa. 1983. *La Agricultura Chinampera*. Mexico City: Universidad Autonoma Chapingo.

———. 1987. "Chinampas of the Valley of Mexico." Paper presented at the Traditional Wetfield Agriculture in the American and Asian Tropics Conference, University of California, Riverside.

Romero, Claudia, and Germán Andrade. 2004. "International Conservation Organizations and the Fate of Local Tropical Forest Conservation Initiatives." *Conservation Biology* 18, no. 2:578–80.

Rosales González, Margarita. 1988. *Oxcutzcab, Yucatán, 1900–1960: Campesinos, cambio agrícola y mercado.* Mérida, Mexico: Instituto Nacional de Antropología e Historia.

Rosenberg, Tina. 2003. "Why Mexico's Small Corn Farmers Go Hungry." *New York Times Online,* www.nytimes.com, accessed March 3, 2003.

Roys, Ralph L. 1965. *Ritual of the Bacabs.* Norman: University of Oklahoma Press.

———. 1976 [1931]. *The Ethno-Botany of the Maya.* Ed. Sheila Cosminsky. 2d ed. Philadelphia: Institute for the Study of Human Issues.

Ruddle, Kenneth, and Ray Chesterfield. 1977. *Education for Traditional Food Procurement in the Orinoco Delta.* Berkeley: University of California Press.

Rugeley, Terry. 1996. *Yucatán's Maya Peasantry and the Ancestors of the Caste War.* Austin: University of Texas Press.

Rugeley, Terry, ed. 2001. *Maya Wars.* Norman: University of Oklahoma Press.

Sabloff, Jeremy A., and William L. Rathje. 1975. The Rise of a Maya Merchant Class. *Scientific American* 233, no. 4:73–82.

Safina, Carl. 1997. *Song for the Blue Ocean.* New York: Henry Holt.

de Sahagun, B. 1950–1969. *Florentine Codex.* Trans. Arthur Anderson and Charles E. Dibble. Salt Lake City: University of Utah Press.

Sanabria, Olga Lucia. 1986. *El uso y manejo forestal en la comunidad de Xul, en el sur de Yucatán.* Xalapa, Mexico: Instituto Nacional de Investigaciones sobre Recursos Bioticos.

San Buenaventura, Fray Joseph de. 1994. *Historias de la conquista del Mayab, 1511–1697.* Ed. and intro. by Gabriela Solis Robleda and Pedro Bracamonte y Sosa. Mérida, Mexico: Facultad de Ciencias Antropologicas, Universidad Autónoma de Yucatán.

Santley, Robert S., Thomas Killion, and Mark Lycett. 1986. "On the Maya Collapse." *Journal of Anthropological Research* 42, no. 2:123–60.

Sauer, Carl. 1952. *Agricultural Origins and Dispersals.* Berkeley: University of California Press.

Schele, Linda, and David Freidel. 1990. *A Forest of Kings: The Untold Story of the Ancient Maya.* New York: William Morrow.

Schele, Linda, and Peter Mathews. *The Code of Kings.* New York: Scribner's.

Scholes, Franz, and Ralph Roys. 1968. *The Maya Chontal Indians of Acalan-Tixchel.* Norman: University of Oklahoma.

Schultz, Theodore. 1964. *Transforming Traditional Agriculture.* New Haven, Conn.: Yale University Press.

Schwartz, Norman. 1990. *Forest Society.* Philadelphia: University of Pennsylvania Press.

Scott, James. 1985. *Weapons of the Weak*. New Haven, Conn.: Yale University Press.

———. 1998. *Seeing Like a State*. New Haven, Conn.: Yale University Press.

Secretaría de Educación Pública (SEP). 1996. *Maaya T'aan*. 6 vols. Mexico: SEP.

Sharer, Robert. 1994. *The Ancient Maya*. 5th ed. Palo Alto, Calif.: Stanford University Press.

———. 1996. *Daily Life in Maya Civilization*. Westport, Conn.: Greenwood Press.

Sheets, Payson. 1998. "Commoner-Elite Interaction in the Southern Maya Frontier: The View from Ceren." Presentation at the Ancient and Modern Maya Symposium, Archaeological Institute of America (Orange County), Irvine, Calif., January 10.

Sheets, Payson, and Michelle Woodward. 1997. "Zoned Biodiversity in Agricultural Fields, Kitchen and Special Gardens, and the Classic Period Landscape." Paper presented at the Annual Meeting of the American Anthropological Association, Washington, D.C.

Shiva, Vandana. 1997. *Biopiracy: The Plunder of Nature and Knowledge*. Boston: South End Press.

Sheridan, Thomas. 1995. *Where the Dove Calls*. Tucson: University of Arizona Press.

Shriar, Avrum J. 2001. "The Dynamics of Agricultural Intensification and Resource Conservation in the Buffer Zone of the Maya Biosphere Reserve, Petén, Guatemala." *Human Ecology* 29:27–48.

Siemens, Alfred. 1998. *A Favored Place*. Austin: University of Texas Press.

Simon, Joel. 1997. *Endangered Mexico*. San Francisco: Sierra Club.

Smil, Vaclav. 1984. *The Bad Earth*. Armonk, N.Y.: M. E. Sharpe.

———. 1993. *China's Environmental Crisis: An Inquiry into the Limits of National Development*. Armonk, N.Y.: M. E. Sharpe.

Smith, Adam. 1910 (1776). *An Inquiry Into the Nature and Causes of the Wealth of Nations*. New York: E. P. Dutton.

Smith, Carol A., ed. 1976. *Regional Analysis*. New York: Academic Press.

Snook, Laura. 1998. "Sustaining Harvests of Mahogany (*Swietenia macrophylla* King) from Mexico's Yucatán Forests: Past, Present, and Future." In *Timber, Tourists, and Temples: Conservation and Development in the Maya Forest of Belize, Guatemala, and Mexico*, ed. Richard Primack, David Bray, Hugo A. Galletti, and Ismael Ponciano, 61–80. Washington, D.C.: Island Press.

Snook, Laura K., and Amanda Barrera de Jorgensen, eds. 1994. *Madera, chicle, caza y milpa: Contribuciones al manejo integral de las selvas de Quintana Roo, Mexico*. Chetumal: ECOSUR.

Sosa, Juan. 1985. "The Maya Sky, the Maya World." Ph.D. dissertation, Department of Anthropology, State University of New York at Albany.

Sosa, Victoria, and J. Salvador Flores. 1993. *La flora ornamental de Mérida*. Mérida, Mexico: Universidad Autónoma de Yucatán.

Sosa, V., Flores, J. Salvador, V. Rico-Gray, Rafael Lira, and J. J. Ortiz. 1985. *Lista FLoristica y Sinonimia Maya*. Etnoflora Yucatánense 1. Xalapa: Instituto Nacional de Investigaciones sobre Recursos Bioticos.

Spencer, J. E. 1966. *Shifting Cultivation in Southeastern Asia*. Berkeley: University of California Press.

Sponsel, Leslie, T. Headland, and R. Bailey, eds. 1996. *Tropical Deforestation*. New York: Columbia University Press.

Stadelman, Raymond (and anonymous editors). 1940. *Maize Cultivation in Northwestern Guatemala*. Washington, D.C.: Carnegie Institution.

Steggerda, Morris. 1943. *Some Ethnological Data Concerning One Hundred Yucatán Plants*. Bureau of American Ethnology, Bulletin 136, Anthropological Paper 29. Washington, D.C.: Smithsonian Institution.

Stephens, John L. 1963 [1843]. *Incidents of Travel in Yucatán*. New York: Dover.

Stevens, Stan. 1997. *Conservation through Cultural Survival: Indigenous Peoples and Protected Areas*. Washington, D.C.: Island Press.

Steward, Julian. 1955. *Theory of Culture Change*. Urbana: University of Illinois Press.

Stiglitz, Joseph. 2003. *Globalization and Its Discontents*. 2d ed. New York: W. W. Norton.

Stonich, Susan C. 1993. *"I Am Destroying the Land!" The Political Ecology of Poverty and Environmental Destruction in Honduras*. Boulder, Colo.: Westview.

Stuart, James. 1993. "Contribution of Dooryard Gardens to Contemporary Yucatecan Maya Subsistence." *Biotica, Nueva Epoca*, 1:53–62.

Stross, Brian. 1973. "Acquisition of Botanical Terminology by Tzeltal Children." In *Meaning in Mayan Languages: Ethnolinguistic Studies*, ed. Munro Edmonson, 107–42. The Hague: Mouton.

Sullivan, Paul. 1989. *Unfinished Conversations*. Berkeley: University of California Press.

Sullivan, Thelma. 1986. "A Scattering of Jades: The Words of the Aztec Elders." In *Symbol and Meaning beyond the Closed Community*, ed. Gary Gossen, 9–18. Albany: State University of New York.

Swezey, Sean, and Robert Heizer. 1977. "Ritual Management of Salmonid Fish Resources in California." *Journal of California Anthropology* 4:1:6–29.

Taube, Karl. 1989a. "A Classic Maya Entomological Observation." *Mesoamerica* (Mérida, Yucatán), summer 1989:13–17.

————. 1989b. "The Maize Tamale in Classic Maya Diet, Epigraphy, and Art." *American Antiquity* 54, no. 1:31–51.

————. 1992. *The Major Gods of Ancient Yucatán*. Washington, D.C.: Dumbarton Oaks Research Library.

Terán, Silvia. 1994. *La Platería en Yucatán*, 2d ed. Mérida, Mexico: La Casa de los Artesanías

Terán, Silvia, and Christian Rasmussen. 1981. *Artesanías de Yucatán*. Mérida, Mexico: Dirección General de Culturas Populares/SEP.

————. 1994. *La milpa de los Mayas*. Mérida, Mexico: Silvia Terán and Christian Rasmussen.

Terán, Silvia, Christian Rasmussen, and Olivio May Cauich. 1998. *Las plantas de la milpa entre los Mayas*. Mérida, Mexico: Silvia Terán and Christian Rasmussen.

Terborgh, John. 1999. *Requiem for Nature*. Washington, D.C.: Island Press.

Thompson, J. E. S. 1971. *Maya History and Religion*. Norman: University of Oklahoma Press.

Tolstoy, Leo. 1958. *Tolstoy's Tales of Courage and Conflict*. Ed. Charles Neider. Garden City, N.Y.: Hanover House.

————. 1962. *Fables and Fairy Tales*. New York: New American Library.

Tuan, Yi-Fu. 1990. *Topophilia: A Study of Environmental Perception, Attitudes, and Values*. New York: Columbia University Press.

Tucker, Mary Evelyn, and John Grim, eds. 1994. *Worldviews and Ecology: Religion, Philosophy, and the Environment*. Cranbury, N.J.: Associated University Presses.

Turner, John Kenneth. 1911. *Barbarous Mexico*. New York: Cassell.

Turner, Jonathan, and Alexandra Maryanski. 1979. *Functionalism*. Menlo Park, Calif.: Benjamin Cummings.

Tuxhill, John, José Luis Chávez-Servia, Jaime Canul Ku, Luis Búrgos May, Victor Interian Ku, and Devra Jarvis. 2003. "Environmental Variability, Farmer Experimentation, and the Conservation of Crop Diversity in Mayan Milpas of Yucatán, Mexico." Paper presented at the Annual Meeting of the Society for Economic Botany, Tucson, Ariz.

UNICEF (United Nations Children's Fund). 2002. *The State of the World's Children 2003*. New York: UNICEF.

Vallarta Velez, Luz del Carmen. 1997. "De la creación de una identidad nacional a través de la religión: La Compañía de Jesús en la frontera México-Belice." Paper presented at the 49th International Congress of Americanists, Quito, Ecuador.

Vargas Rivero, Carola A. 1983. "El *ka'anche':* Una practica horticola Maya." *Biotica* 8, no. 2:151–73.

Várguez Pasos, Luis, ed. 1981. *La milpa entre los Mayas de Yucatán.* Mérida, Mexico: Universidad Autónoma de Yucatán.

Varner, John Grier, and Jeannette Johnson Varner. 1983. *Dogs of the Conquest.* Norman: University of Oklahoma Press.

Vayda, Andrew, and Bradley Walters. 1999. "Against Political Ecology." *Human Ecology* 27:167–79.

Villanueva Mukul, Eric. 1985. *Crisis henequenera y movimientos campesinos en Yucatán 1966–1983.* Mérida, Mexico: Instituto Nacional de Antropología e Historia.

———. 1990. *La formación de las regiones en la agricultura (El caso de Yucatán).* Mérida, Mexico: Maldonado.

Villa Rojas, Alfonso. 1945. *The Maya of East Central Quintana Roo.* Publication 559. Washington, D.C.: Carnegie Institution.

———. 1978. *Los Elegidos de Dios.* Mexico City: Instituto Nacional Indígenista.

———. 1985. *Estudios Etnologicos: Los Mayas.* Mexico City: Universidad Autónoma de México.

Villaseñor Rios, José Luis, and Francisco J. Espinosa García. 1998. *Catálogo de malezas de México.* Mexico City: Universidad Autónoma de México.

Villers Ruiz, L., R. M. Lopez Franco, and A. Barrera. 1981. "La Unidad de Habitacion Tradicional Campesina y el Manejo de Recursos Bioticos en el Area Maya Yucatánense," Pt. II. *Biotica* 6, no. 3:293–323.

Voeks, Robert. 1997. *Sacred Leaves of Candomble.* Austin: University of Texas Press.

Vogel, Joseph. 1994. *Genes for Sale.* Oxford: Oxford University Press.

———. 2000. *The Biodiversity Cartel.* Quito, Ecuador: CARE.

Wagner, Henry R., ed. 1942. *The Discovery of Yucatán.* Berkeley: Cortes Society.

Warner, Lloyd. 1953. *American Life: Dream and Reality.* Chicago: University of Chicago Press.

Weaver, Nevin, and Elizabeth C. Weaver. 1980. "Beekeeping with the Stingless Bee *Melipona beecheii* by the Yucatecan Maya." *Bee World* 62, no. 1:7–19.

Weber, Max. 1951. *From Max Weber.* Ed. by Hans Gerth and C. Wright Mills. Oxford University Press, New York.

Wechsler, Doug. 1988. "Dark Times for Cuba's Sabal Palms." *International Wildlife* 28, no. 2:38–43.

Webster, David. 2002. *The Fall of the Ancient Maya*. London: Thames and Hudson.

White, Christine D., ed. 1999. *Reconstructing Ancient Maya Diet*. Salt Lake City: University of Utah Press.

Whitmore, Thomas, and B. L. Turner II. 2001. *Cultivated Landscapes of Middle America on the Eve of Conquest*. New York: Oxford University Press.

Wilken, Gene. 1987. *Good Farmers*. Berkeley: University of California Press.

Wilk, R. R. 1981. "Agricultural Ecology and Domestic Organization among the Kekchi Maya." Ph.D. dissertation, University of Arizona.

————. 1991. *Household Ecology*. Tucson: University of Arizona Press.

Williams, Michael. 2003. *Deforesting the Earth*. Chicago: University of Chicago Press.

Williams, Nancy, and Eugene Hunn, eds. 1982. *Resource Managers*. Boulder, Colo.: Westview.

Wilson, David Sloan. 1998. "Hunting, Sharing, and Multilevel Selection: The Tolerated-Theft Model Revisited." *Current Anthropology* 39:73–99.

Wrangham, Richard, and Dale Peterson. 1996. *Demonic Males*. New York: Houghton Mifflin.

Wright, Angus. 1990. *The Death of Ramon Gonzalez*. Berkeley: University of California Press.

Xiu C., Gaspar Antonio. 1986. *Usos y Costumbres de los Indios de Yucatán*. Mérida, Mexico: Maldonado.

Young, James. 1981. *Medical Choice in a Mexican Village*. New Brunswick, N.J.: Rutgers University Press.

Zaid, Gabriel. 1995. *Omnibus de poesía mexicana*. 18th ed. Mexico City: Siglo Veintiuno Editores.

Zavala, M. 1974 [1896]. *Gramática Maya*. Mérida, Mexico: José Díaz Bolio.

Index

About the Authors

E. N. Anderson is Professor of Anthropology at the University of California, Riverside. His major interest is in the area of cultural and political ecology, with a focus on how people understand and use plant and animal resources. He has done research in Hong Kong, Malaysia, British Columbia, Mexico, and elsewhere. His publications include *The Food of China* (Yale University Press, 1988); *Ecologies of the Heart* (Oxford University Press, 1996); *A Soup for the Qan* with Paul Buell and Charles Perry (Kegan Paul International, 2001); and *Animals and the Maya in Southeast Mexico* with Felix Medina Tzuc (University of Arizona Press, 2005).

Felix Medina Tzuc was born in rural Yucatán state, Mexico, in 1942. As a young man he moved to the area of Chunhuhub, then largely a trackless old-growth forest. He married Elide Uh May, and they spent most of their lives subsistence farming on an isolated plot of land south of Chunhuhub. Eventually they moved to town, planting orange trees on their parcel of land. Don Felix has served a term as justice of the peace for Chunhuhub. He has a biologist's eye for the natural world and has devoted a lifetime to learning about plants, animals, farming, and field skills in Quintana Roo's tropical forests.

Pastor Valdez Chale is a bilingual education specialist (Spanish/Maya) living in Chunhuhub and serving as principal of the elementary school in Presidente Juarez. He is married to Zenaida Estrella; they have six children.

Aurora Dzib Xihum de Cen lives in Presidente Juarez, where her family carries out subsistence agriculture. She is studying to be a teacher. She and her husband, Jaime Cen, have one son.